American Nuclear Deception: Why "the Port Chicago experiment" must be investigated

Daisy B. Herndon

Published by The Port Chicago Witness, 2022.

AMERICAN NUCLEAR DECEPTION: WHY "THE PORT CHICAGO EXPERIMENT" MUST BE INVESTIGATED

First edition. August 29, 2022.

ISBN: 979-8986053615

Written by Daisy B. Herndon.

Table of Contents

Preface 1

A thin and careful veil 7

How I learned to stop assuming I knew history and started wondering about the bomb 15

Person of interest 22

Truth, lies and consequences 24

An outlandish theory 27

The power of the past 34

Evidence 35

Terms to know 40

PCnet people 51

The story of the bomb story 64

The Port Chicago nuclear explosion theory (PCnet) 67

The Port Chicago / Manhattan Project connection 70

"In typical Port Chicago fashion" 72

"A Story Too Good to Kill" 75

Overcoming the objections 80

See no evil 88

Seeing the big picture 90

Why would there have been a test at Port Chicago? 91

Why the PCnet matters today 94

The "Einstein letter" story 105
A providential opportunity 113
The truth about Truman 115
The ongoing "atomic crusade" 123
Operation Crossroads 125
The first atomic admiral 131
Racism, pragmatism and the PCnet 134
"BuNav . . . invent something" 142
Necessary ruthlessness 156
"The source of his power" 161
The quest for a nuclear torpedo 166
Parsons at Port Chicago 171
The disappearing Thin Man 177
The Port Chicago explosion and the Manhattan Project 189
Designed for large explosions 194
An accident designed to happen 197
Deak Parsons, cowpuncher 202
To see what he could see 206
Research 208
Improbable cause 210
Blame without a cause 212
Coincidences? or coordinated incidents? 217

Discredit? or distraction? 222
Focus and perspective 232
Comparable effects 236
The PCnet and democracy 238
FDR's July journey 240
The President's flagship 249
FDR's ongoing health problems 252
Effects, and causes 258
Einstein's torpedo 260
An intricate network 263
America's bomb 267
Reviewing Einstein's well-kept secrets 272
"A single bomb, carried by boat..." 277
Parsons' power 282
FDR's big atomic nightstick 284
Race and the "race for the bomb" 289
What if... 292
Evidence pertaining to the PCnet, a review 294
Verdict? 311
Acknowledgements: 312
Dedication 313

To my grandchildren

"Today we are faced with the preeminent fact that, if civilization is to survive, we must cultivate the science of human relationships—the ability of all peoples, of all kinds, to live together and work together, in the same world, at peace."

Franklin D. Roosevelt, April 13, 1945, Final speech (*undelivered*)[1]

"The scientific revolution that led to the splitting of an atom requires a moral revolution as well."

Barack Obama, Hiroshima speech, May 27, 2016[2]

Preface

There was a bomb shelter in the basement of the house where I grew up in Pontiac, Michigan, a factory town 30 miles north of Detroit. Bomb shelter, tornado shelter, pantry—a small, multipurpose room. Its walls were concrete blocks, its floor covered with gravel. Rows of Mason jars lined the dusty shelves that spanned the back wall. Canned peaches, canned tomatoes, canned green beans . . . Daddy's garden and Mommy's hard work helped keep their twelve children well-fed and healthy.

Our family never talked about the 'Duck and Cover' exercises we did at school. When the teachers said "Duck," we ducked—filed into the hallway and crouched in a row, our heads tucked against the wall. Safe, I guess, from something. (I wonder, now, if that was where we got the idea for the game of tag we called "Duck, Duck, Goose." I could Google it, I know. But I won't. We have other important things to look into right now.)

Ordinary Americans seldom talk about nuclear weapons except when an imminent threat of nuclear war makes headline news. On a routine basis, most of us pay little attention to the ongoing problems millions of others face as a result of nuclear fallout and mismanagement of nuclear wastes.

Yet those hot topics are everyday concerns, not only for anti-nuke activists, scholars and pundits, but for the millions of ordinary Americans who are members of groups that have been direct victims of radioactive fallout—Native American 'yellowcake' miners; the Atomic Veterans and their families; descendants of the various Downwinders groups; residents of St. Louis, Missouri; and other unwitting civilians.

Studies show that radioactive fallout from the Trinity test may have impacted infant mortality rates among residents within a thousand-mile radius of the blast. According to **The Bulletin of the Atomic Scientists,**

> "The 21 kiloton explosion occurred on a tower 100 feet from the ground and has been likened to a "dirty bomb" that cast large amounts of heavily contaminated soil and debris—containing 80 percent of the bomb's plutonium—over thousands of square-miles."[3]

Even more surprising, the Trinity fallout spread across the country. The "delayed or long-distance fallout" was only detected after it caused defects to appear on film at the Eastman Kodak Company in New York. According to recent estimates, fallout from decades of nuclear testing between 1951 and 1973 may have been responsible for the deaths of up to 690,000 Americans.

And, of course, there is the ever-looming uncertainty that results from the fact that our security—not only of our nation but of the world—depends on the wisdom and sole discretion of powerful and independent leaders of nuclear nations. As we know, some of them have found the "big button" rather irresistible.[4]

And then there is the **Port Chicago nuclear explosion theory**.

This book tells what I learned when I looked into the strange allegation concerning the little-known but massive blast that took place on July 17, 1944, one full year before the famous Trinity test. In the following pages, you will learn how an ordinary citizen, Peter Vogel, an information officer working for the State of New Mexico, came up with the outlandish *and yet increasingly plausible* idea that the devastating explosion that killed 320 Americans at the Port Chicago Naval Ammunition Depot near San Francisco, California may have been the world's first secret nuclear test.

At this point, you are probably rolling your eyes, as most people do when they first hear the "ridiculous" claim. As I did—before I discovered the evidence presented here.

Believe it or not, the theory won't just go away. Though deleted from *Wikipedia*, it still shows up in several sources, including scandalous posts on the internet.

In recent years, social media has helped promulgate all kinds of wild (and sometimes dangerous) conspiracy theories. Who and what are we to believe? There are so many people (yes, including me) reporting "news" and "information." Telling the truth is optional. Fact-checking is not a requirement. So how do we discern between valid information and 'hogwash'?

Sometimes when stuff sounds ridiculous, the best thing we can do is apply simple logic—that is, figure out what makes sense, using our best judgment, based on what we know. To assess rumors that Sasquatch is stomping around town, such reasoning may suffice.

On the other hand, both the National Archives and the U.S. Navy thought it worthwhile to investigate the story of the so-called "Philadelphia Experiment". After due diligence, each agency concluded that there is no scientific basis for believing that a Naval ship disappeared in the fall of 1943. We can safely assume, then, that they have permanently disproved the theory that, while World War II was raging overseas, and unwitting Americans were hard at work secretly building a weapon that would help shift the world order, crafty engineers tested their teleporting skills by secretly sending an invisible ship from the Philadelphia Navy Yard to the Tidewater area and back.

(Deeper investigation reveals that the Navy was indeed looking into a process called degaussing as a way to shield warships from detection by radar. And that may be the source of the fire behind that smokescreen.)

In December 1999, the Army reported the negative results of its 16-month investigation into persistent and growing rumors that 1200 African American soldiers had been massacred at a Mississippi army base in 1943. It is no secret that the war years were hard on Black military personnel, as historian Matthew F. Delmont substantiates in his 2022 book, **Half**

Americans[5]. Black people in the military were fighting for "double victory", for freedom at home as well as overseas. According to the *Washington Post,* the rumors of a massacre at Camp van Dorn may have grown from exaggerated reports of real incidents. The Army study revealed multiple problems, including

> ". . . inadequate housing, training and recreation facilities and, more seriously, two shootings of soldiers, one by a small-town Mississippi sheriff and the other by military police, that prompted riots by black soldiers."[6]

When the Army's report was released, leaders of the NAACP responded by asking then-Attorney General Janet Reno to investigate the investigation. John White, a spokesman for the NAACP, told the *Washington Post,* "We just feel that these charges are too serious for us to say, 'Great. The Army's correct. Let's move on."

Sailors at Port Chicago faced problems similar to those experienced by Black soldiers throughout the country during World War II. That partially explains why the massive disaster—one of the largest man-made explosions in history—became the backdrop of the largest mass court martial in U.S. Naval history.

In the aftermath of the explosion, fifty Black sailors were found guilty of mutiny after refusing to load ammunition onto ships. Many people who know the story assume they were protesting racial injustice and poor working conditions; however, their defense was based on their expressed fear of another unexplained explosion.

Their case became a public cause, thanks in part to the involvement of the NAACP, and because the Navy made the uncharacteristic decision to publicize the military tribunal. After the war, public outcry over the injustice helped bring about the integration of the Navy. Decades later, Dr. Robert Allen revived the story with his book, **The Port Chicago Mutiny**, which in turn led to creation of the Port Chicago Naval National Memorial, where an annual service commemorates the sacrifice of those who died in the blast. The renewed attention also spurred ongoing legislation to exonerate "The Port Chicago 50."

And then there is the PCnet.

What is easily overlooked in the story of Port Chicago is the fact that the mass mutiny trial upstaged the story of the massive explosion. If my updated version of the PCnet is accurate, that diversion may have been by design. (After all, what is a good conspiracy theory without intricate devious scheming? But if there is merit to this suggestion, it must be supported by evidence. Read on.)

The PCnet is not your granddaddy's conspiracy theory. As quiet as it has been kept, the historical allegation is part of nuclear history. After all, if there was a nuclear test at Port Chicago, it was conducted under the auspices of the top-secret bomb project. Unlike the so-called 'Philadelphia experiment', the Port Chicago nuclear explosion theory has never been debunked—because the logistically feasible claim has never been investigated. Which makes sense, especially considering that the astonishing success of the bomb, augmented by an effective public information campaign, helped Manhattan Project leaders avoid an in-depth Congressional review of the secret project after the war.

The deliberate steps they took to limit transparency (yes, an intentional euphemism for 'cover-up') has lasting consequences. It bears repeating throughout this book: we still do now know what we never learned about America's bomb project.

A thin and careful veil

When Manhattan Project leaders decided what the public should "know", they were concerned not only about the immediate reaction to the bomb but also about how the story would be told in years to come. The slow demise of the *Thin Man* bomb is one alarming result that would probably go unnoticed but for the persistent story of the PCnet.

The plutonium gun-assembly weapon "shelved" by Manhattan Project leaders (on the day of the Port Chicago explosion) was the world's first atomic weapon. Yet there are no museum displays of the prototype, which is increasingly omitted from historical accounts. Why?

Twenty minutes into the *History Channel*'s documentary, **Hiroshima: 75 Years Later**,[7] the *Thin Man* bomb is obliquely referenced (with misleading graphics) as "the design that was then current." You only recognize the veiled reference if you already know the paradigm-shifting story. Yet the disappearing *Thin Man* has unique and profound historical significance. After spending the first year of the three-year project developing the prototype weapon, the decision to shelve it caused a complete reorganization of the Los Alamos Laboratory.

The problem with the *Thin Man* was known even before the Lab was established. Most histories do not explain why scientists *and engineers* continued working on it anyway: they believed the plutonium gun-assembly model would at least provide proof of the uranium gun. The *Little Boy* was the successor to the *Thin Man*.

Without that tidbit of information, we accept the oversimplified "fact" that there was allegedly "no test" of the *Little Boy* bomb before it was dropped on Hiroshima. Why there was no test of the *Little Boy* is a question that is repeatedly met with elaborate and unsubstantiated reasoning that defies both logic and precedent.[8]

But what if the *Thin Man* did indeed provide proof of the *Little Boy*? If so, what was the proof of the *Thin Man*? As we will see, evidence supporting the PCnet raises these previously unasked questions, and more.

History, we must remember, is not the full story of exactly what happened. It is the tale certain people weave from records kept by others. And many untold stories result from the absence of adequate records.

We erect monuments, write books, build museums, and create National Archives to preserve important details of history, the details that shape our national identity. When someone fails to record or wipes out historical facts, they do so with consequence. After all, omission, deletion and misrepresentation of information are some of the most standard and effective ways to keep a secret or to shape a narrative.

We are all accustomed to nuclear secrecy, which, we assume, was essential to national defense. We accept without judgment the stories of the secret cities with their cult-like isolation and security. Some 600,000 Americans worked directly for the bomb project at some point between 1942 and 1945 without knowing what they were working on. (Imagine that happening today!)

The President managed to fund the war project without letting Congress in on the scheme. Yet today, it just seems downright un-American to suspect that the classified files at the Los Alamos National Library and the Department of Energy may hold unsavory secrets. Or that, as other evidence reveals, some secrets were too hush-hush to have ever been committed to paper.

Oh, rest easy. There are no security breaches in this book. I did most of the research online. I accessed primary documents from public archives and libraries, and studied secondary sources readily available to the general public. No security clearance necessary (or desired.)

The Port Chicago nuclear explosion theory did not surface until the 1980s, when Vogel stumbled upon the document that changed his life—and may yet change history.

There really is something to be said for hindsight. This book will show how and why we can use an unusual idea like the PCnet to gain profound and surprising insight into unanswered historical questions—including some we didn't even know enough to ask. And, the revealing theory shows why cherishing and maintaining our curiosity is the civic duty of citizens in a democracy.

Ask we must; because a nation is—at best—the deprived orphan of its unknown history. At worst, we are like the Trinity Downwinders who were told flat-out lies about the radioactivity they were exposed to. (General Groves issued a false press release stating that there had been an explosion at a remote ammunition magazine. Hmm. Wonder where he got the idea that a massive explosion of conventional ammunition would make a good cover story for a nuclear blast.)

In many ways, we are all victims of the bomb, with its many underreported casualties and its exorbitant costs. Perhaps more importantly, we are victims of the false narrative of the bomb. The only nuclear weapons ever used in a war were those that fell on Japan; at the same time, however, the nuclear age fell on the whole world.

As citizens of the democratic nation responsible for creating and using the weapon, Americans are custodians of the hidden history, a false narrative that may be as hazardous for democracy, and for humanity, as the carelessly stored plutonium that devastated the lives of so many unsuspecting residents of St. Louis, Missouri.[9]

Nuclear history is important to everyone. The widely accepted premises of nuclear history, including false or unproven assumptions, contribute to ongoing nuclear policy. World leaders who rely on this body of 'knowledge' make decisions about our lives. This is one of the reasons why we all need to talk more about nuclear weapons, why we need to ask questions and insist on true answers. In addition to the ever-looming threat of nuclear war, there is much more to talk about; much more we need to know.

For a start, we need to be clear about the fact that **we do not know the real history of the bomb.** And what we don't know can kill us. Hopefully, though, there is truly freedom to be gained from knowing the truth. Knowing, not assuming.

The familiar story of the bomb was based on the narrative created by the stakeholders, people like Manhattan Project Director General Leslie R. Grove, Secretary of War Henry Stimson, and others involved in the creation of the weapon. As you will see from clips in this book, they wanted the public to 'know' that the war-winning weapon saved lives, in a cost-effective way. Their version of the *classified* story was the only one available to the journalists who wrote the first draft of nuclear history.

That slanted narrative was the basis of public information about the bomb until the 1960s, when declassified documents first became available. Historians began reassessing the story based on the new information that emerged in that era, but it changed the story only for those willing to consider the new interpretations.

Today, the 'revisionist history' of the bomb features the idea of "atomic diplomacy", the claim that the bomb was used against Japan as a means of sending a signal to Russian Premier Joseph Stalin. That assessment is hotly debated by those who cling to the traditional narrative.

But neither wave of nuclear history has been satisfactory. Although new information allowed for new interpretations, the basic premises were still built upon the original faulty foundation, which was adopted with little scrutiny.

One key example is the so-called "Einstein letter"—better known today as the "Szilard-Einstein letter" now that historians recognize physicist Leo

Szilard as the real author. Historians still overlook the important role of Alexander Sachs, the Wall Street banker who reportedly delivered the letter to President Roosevelt.

Nor did the change of acknowledged authors lead to a review of the basic premises. So, historical accounts continue to interpret Einstein's letter as a warning to FDR about a German bomb. (The letter's explicit objective was to obtain Government support for increased uranium research.)

Although historians generally agree that the letter was not the catalyst for the U.S. bomb project that evolved three years later, that faulty premise also remains part of the accepted rationale for creation of the bomb. This is not just in the 'old school' eye of the public, but in the story told by 'revisionists' as well. Ever-revolving and unresolvable debates about the decision to use the weapon rest upon that foregone and faulty conclusion.

Like the original narrative, revisionist accounts have relied on ungrounded assumptions. As a consequence, there are still many unanswered questions. So today, historians are engaged in what some call a "nuclear history renaissance", once again seeking fresh information to update the story of the bomb. In this third go-round, researchers should leave no stone unturned.

The PCnet provides an unusual tool for examining various un- or under-explored aspects of nuclear history. It gives new insight, for example, into President Franklin Roosevelt's unilateral and secret decision to create the weapon. Until recently, that was one of the important skirted by historians, who have been thwarted by FDR's notorious secrecy. Taking advantage of the PCnet will help researchers get a fuller picture. The theory offers a unique point of view that addresses many unanswered questions and indicates new areas of research.

The takeaways are practical and timely. If there is one thing we can learn by studying the history of the bomb, it is that a small handful of empowered people can change the world in dynamic ways. For better or for worse. With or without the knowledge and participation of citizens. An informed public is essential to democracy. Information is empowering. It can be used to make things worse, or it can be used for better. But for the truth to set us free, we need to know the truth.

I was repeatedly surprised by what I leaned about the bomb through the lens of the Port Chicago nuclear explosion theory. Readers who open their minds will be astonished by the eye-opening evidence. And enlightened. (And, perhaps a bit outraged, but not to worry; righteous citizen indignation can be channeled into positive civic action.)

Again, despite what some people think they know, **the Port Chicago nuclear explosion theory is a plausible claim that has never been debunked.** As the evidence clearly shows, it was logistically quite possible to carry out a clandestine test at the remote military base. The powerful and secretive people who could have carried out the experiment had sufficient motive and opportunity to do so. And it is clear that when it came to exposing its own citizens to nuclear hazards, the U.S. Government has exhibited callous conduct in the past.

Yet there has been no official investigation of the increasingly important claim. Still, most people who have heard of the PCnet dismiss it as a "mere conspiracy theory", substituting their typical opinionated logic for evidence.

"*Where's the radioactivity?*" they ask—as though the question is its own answer. It is not. For one thing, evidence of radioactivity is a complex topic, and, as is always the case, the apparent lack of evidence is not evidence of lack. Besides, what do we do with the information that radioactivity from nuclear testing has shown up across the U.S., even in our milk[10] and honey[11]?

"*We would have known by now*", they declare—as though we could know the facts without having ever pursued them.

"A highly populated military base would be an illogical choice for a test," they assert; as if they have studied either the history of Port Chicago—not the Port of Chicago[12]; a distinction that confuses some people—or the criteria

for secret nuclear tests. This objection demonstrates how little people know about the early U.S. nuclear tests that did indeed put both military personnel and civilians at risk.

"*There was not enough uranium or plutonium for a nuclear bomb*", they argue—as if an experiment designed to obtain data for use in creating the bomb would have required as much material as a full-scale weapon.

"*We would have built the bomb sooner and used it against Germany,*" they insist—clearly unaware of the historic documents that say (and show why) Japan "was always the target".

So much for a 'mere conspiracy theory'.

Without a thorough investigation—without seeking, gathering, examining and analyzing the evidence—all conclusions about the PCnet are mere assumptions. Of the many things to be learned from this story, one important takeaway is that when it comes to evaluating unfamiliar ideas, logic beats opinion and evidence trumps logic.

In one forum or another, the PCnet continues to surface. Government sources, historians and history buffs prove nothing by dismissing the uninvestigated theory as 'ridiculous', 'hogwash. To prove or disprove an outrageous but logistically possible, *research-based* claim like the PCnet, sound evidence is a requisite. After all, even the historians who provided the harshest and most in-depth criticism of Vogel's theory admitted that he asked the right questions and accessed good sources. For the most part, they did not even say his evidence was wrong; they just disagreed with his conclusions.

Whether or not the theory itself withstands close scrutiny, the Port Chicago nuclear explosion theory warrants careful investigation. The clear but underreported link to the Manhattan Project is enough to give the explosion unique historical significance.

The explosion played a vital role in the creation of the world-changing weapon; that should be the single most defining feature of the Port Chicago story. The disaster—which, again, is best known as the story of the mass mutiny trial—played a notable role in African American history, but the massive explosion was crucial to the development of the bomb. This important detail has been omitted from accounts of both Port Chicago and the Manhattan Project, an omission that lends credence to the nuclear explosion theory. After all, withholding significant information was one of the primary means of keeping the Manhattan Project secret.

Of course, we only discover this little-known fact by conducting an investigation. Up till now, such investigation has been barred by assumptions. In July 1944, the Navy board that investigated the explosion averted further investigation by ruling that we can never know the cause. Today, many people jump to the logical (and handy) conclusion that it was an accident, unaware that they are basing their opinion on the flawed history, not on the evidence. They will be astonished by the many unknown and important historical facts that are revealed through the lens of the PCnet.

How I learned to stop assuming I knew history and started wondering about the bomb

In the late 1990s, I was conducting research for a novel set in World War II when I read about the massive explosion at the Port Chicago Naval Ammunition Depot near San Francisco, California on July 17, 1944, where 202 of the 320 Americans killed were Negro sailors loading ammunition onto warships bound for the war in the South Pacific. I decided Port Chicago would be a perfect setting for the story of a young African American youth who joined the Navy to prove his mettle. I decided to make him a survivor of the historic blast.

FIRST with the NEWS

The New York Age

5c

200 Negroes Among Calif. Explosion Dead

Wadleigh, Benjamin Franklin High Schools To Continue Operations; Full Supervisory Staffs Retained

TWO HUNDRED NEGRO SAILORS AMONG DEAD, MISSING AND INJURED IN CALIF. EXPLOSION

I was coming right along with my 'Great American Novel', **Scrap Mettle**, when I stumbled upon the Spring 1982 edition of **The Black Scholar**. The cover story was Peter Vogel's surprising article, "*The Port Chicago Disaster – Was it a Nuclear Explosion?*" The issue was edited by Dr. Robert Allen, author of **The Port Chicago Mutiny**.

THEBLACKSCHOLAR
THE PORT CHICAGO DISASTER
Was It A Nuclear Explosion?
200 black sailors died in a huge explosion at a U.S. Navy Base in California in 1944. Were they— and others who died— victims of an unacknowledged nuclear disaster?
Special Investigative Reports in this Issue.

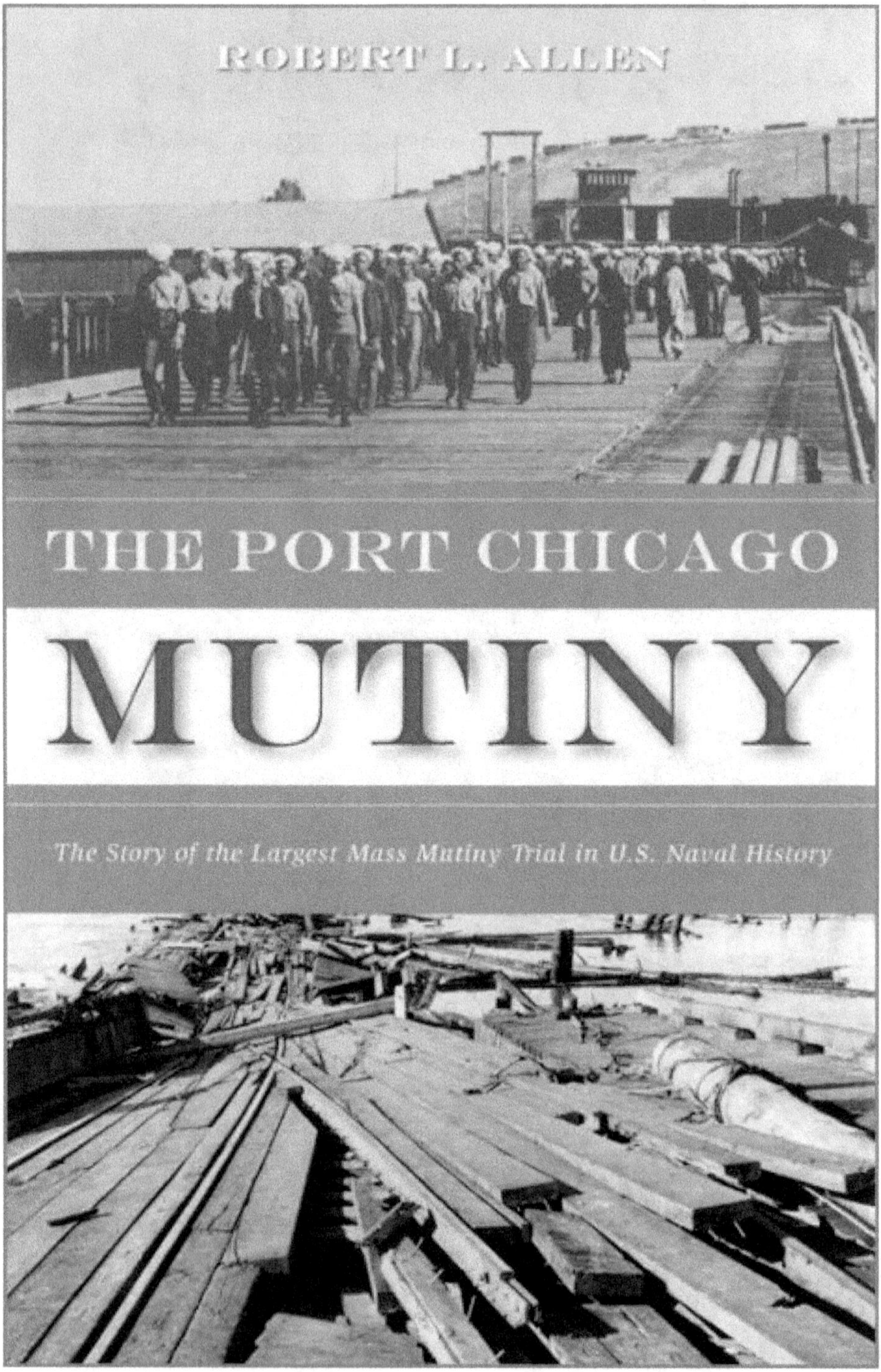
ROBERT L. ALLEN
THE PORT CHICAGO
MUTINY
The Story of the Largest Mass Mutiny Trial in U.S. Naval History

According to Allen, Vogel eventually dropped the theory for lack of evidence. That may have been a temporary decision on Vogel's part, but he continued his grueling, thankless research for over 35 years. He updated his findings in an online eBook, **The Last Wave from Port Chicago**. In December 2018, when he shut down his website and withdrew the book (for personal reasons), the senior independent scholar remained convinced that the Port Chicago explosion was a secret test conducted by the leaders of the Manhattan Project during the research and develop of the atomic bomb.

But it was not until I noticed a peculiar anomaly in the transcript of the Port Chicago mutiny trial that I decided to set aside my work on **Scrap Mettle** just long enough to learn the basic facts about the PCnet. When I first began investigating the theory, I thought I would quickly find out why it was ridiculous. Surprise, surprise, surprise after surprise!

The Port Chicago nuclear explosion theory has a lot in common with two cautionary tales, *The Blind Men and the Elephant* and *The Emperor's New Clothes*. In the first story, each (male) investigator arrogantly assumes he has the whole picture. In the latter, the politically risky truth about the status of the emperor is revealed by a naive kid who apparently had not been told that "loose lips sink ships".

As an African American grandmother, I identify more closely with the loudmouth servant in the Spanish version of the *Emperor's New Clothes*, an 'old negro with nothing to lose'. Unlike him, though, I have a world to gain by speaking up. You see, my 18-year-old granddaughter was recently inducted into the Navy. Tearfully watching the video of her swearing-in ceremony, I renewed my own vow: to do all I can to serve and protect my six grandchildren; to serve my country; to serve humanity and, in doing so, to serve my God. I am doing what we were all instructed to do after 9/11: I see something. That's why I'm saying something.

It may be that historians missed the bigger Port Chicago story simply because the 'conspiracy theory' did not meet academic standards as a legitimate area for inquiry. Or, shrewd professionals may have seen it wise not to wade deeper into the unsanctioned territory buried far left of the beaten narrative. Further investigation, for example, of the historic quest for an atomic torpedo[13] might have come a little too close to what Lee Basham calls "toxic truth".[14]" The South Texas College philosophy professor (who believes conspiracy is bad for democracy) has studied the Port Chicago nuclear explosion theory as an example of how a major secret could be contained for decades if the small and powerful group controlling that information deemed it necessary.

(To be clear, I am in no position to reach any conclusions about why historians have not explored the PCnet. I hope they will, though, in the future. As a non-academic, my self-published work has not been subject to peer review.)

Unlike Basham, some scholars I reached out to have been reluctant to engage, at least so far. By contrast, I was encouraged by the historians who looked over my initial research when I attended the 2018 Conference of the Association for the Study of African American Life and History (ASALH). The theme was "African Americans in Times of War." While few were familiar with the Port Chicago story, they were surprisingly open to the possibility of a nuclear test, and encouraged me to write the story. I have poured myself into this work, endeavoring to make it accurate, well-reasoned and readable. Because I believe the topic is important, I welcome and encourage all close scrutiny and comments.

As I said earlier, I was amazed at the historical information I discovered, mostly in primary sources, that does not appear in accounts of either Port Chicago or the Manhattan Project, nor in any other published source. This book only presents information directly related to the PCnet. I am planning two additional books to explore the evidence from other angles.

I am also developing a proposal for a community research project that will enlist historians, journalists and educators to help **interested citizens** create a crowd-sourced history of the bomb, a well-documented history told by and for the people, presented in as many ways as creative people can imagine. The "*PCW (Port Chicago Witness) Project*" will complement similar initiatives already underway, but here the emphasis will be on getting ordinary people from all walks of life involved in doing the research and publicizing the story.

Because nuclear history is our history. The Port Chicago story is best known as African American history. The PCnet shows one reason why we need to recognize African American history as American history. As Americans, it is our right, it is our responsibility, and we have tremendous reasons to know what the Port Chicago explosion had to do with the Manhattan Project and the creation of the atomic bomb.

We know the disaster was a catalyst for civil rights in the military. And we need to know the rest of the story.

Person of interest

One of the three historical figures of greatest interest to the Port Chicago Experiment is the obscure but powerful man best known as the Navy captain who armed the *Little Boy* bomb in midair as the *Enola Gay* flew its fateful cargo to Hiroshima on the morning of August 6, 1945.

In fact, William S. Parsons was the head of ordnance for the Manhattan Project, the man responsible for producing, testing and delivering the first atomic bombs. After the war, he went on to become the first "atomic admiral." The only man to either witness or participate in seven of the first eight nuclear explosions, Parsons had a lasting influence on U.S. nuclear policy.

After reading his only biography, **Target Hiroshima, Deak Parsons and the Creation of the Atomic Bomb** by Al Christman, one reviewer concluded that "Parsons' career reads like an object lesson in the value and importance of engineering knowledge in scientific and technological development."

Although it borders on hero worship, **Target Hiroshima** is a very informative book. Christman had to acknowledge that Parsons' life "reveals both lessons and cautions concerning the use of science for military purposes."

Christman also points out the advantage of reviewing the history of the bomb from a distance in time.

> "In particular, the radiation hazards and the moral issues of the atomic bomb are seen in a more discerning light now than when still under the shadow of World War II. These issues, plus *revisionist efforts to reshape nuclear history*, make it **all the more important to examine the circumstances that brought the atomic bomb into being** and into combat use."[15] [emphasis added]

"Revisionist" is a derogatory term used by traditionalists to describe anyone who believes there's more to the story than what we already know. But what's the point of historical research if not to update the story?

As the African proverb says, until the lion tells the story, the glory goes to the hunter. In the case of the bomb, the hawkish lions became the narrative-makers. The stated purpose of their post-war propaganda was to sell the bomb to the public and to prevent people from taking a negative attitude toward the bomb:

- The weapon produced by "the greatest achievement of organized science in history."

- The weapon by which a small group of men harnessed "the basic power of the universe."

- The exponential and cost-effective weapon that killed 100,000 people in one blow.

Truth, lies and consequences

History serves a purpose. Right or wrong, true or false, history has an effect. (Where is the account that heralds love as the basic power of the universe? Where is the history that portrays the heart and soul of the people as the basic power of democracy?)

The faulty history of the atomic bomb paved the way to our current nuclear policy. Consider the utter madness of a value system based on a policy of Mutually Assured Destruction, one that prioritizes more or less useless weapons of mass destruction as the best means of national security. This is the psychological fallout that permeates a world where masses of people find it necessary to raise a slogan declaring that "Black Lives Matter" – something that should be a foregone conclusion, and would be if all lives mattered.

It is not insignificant that the most direct victims of the first 70 detonations of nuclear weapons by the United States were people of color.

- An estimated 200,000 Japanese civilians died in the atomic bomb attacks on Hiroshima and Nagasaki in August 1945.

- The people of the Marshall Islands were uprooted from their homeland when the U.S. told the people their "temporary" evacuation was "for the good of all mankind"—and went on to conduct 67 nuclear tests there between 1946 and 1968.

- And, for the last few years, I have been investigating a question featured in the spring 1982 edition of **The Black Scholar**: "*The Port Chicago Disaster - was it a nuclear explosion?*"

A vital test

In July 1942, President Roosevelt wrote a letter to the National Association for the Advancement of Colored People (NAACP) in which he reversed his 1940 decision to maintain a segregated Navy. Rephrasing the slogan of the convention, FDR declared that "Minorities are Vital to Victory." After all, there was a war on. And a world-changing weapon to be produced. And tested. Somewhere.

A present dilemma

In a 2020 book on nuclear history, **The Age of Hiroshima**, seventeen contributors examine the story of the nuclear revolution and contemplate the current nuclear history renaissance. These scholars agree that there are many unanswered questions in the traditional history of the bomb, which, they admit, is grounded in too many faulty assumptions. Some experts also express concern that the topic does not seem to garner sufficient interest among the general public.

Where is *Bert the Turtle* when we need him? We need him now!

"*Duck and Cover*", U.S. Civil Defense campaign, early 1950s

In the meantime, and for a start, perhaps this book will do. After all, there's a new nuclear age on. Can we talk?[16]

You be the judge

In this brief introduction to the Port Chicago nuclear explosion theory (PCnet), my aim is to present the evidence as if each piece was an exhibit in a grand jury hearing. You will see, firsthand, some of the key documents that led me to the conclusion that Port Chicago was designed for nuclear explosions.

Here, I present the evidence that favors the updated theory, but *you, Reader, are the judge*. In the end, it is up to you decide:

> ***Was* the Port Chicago Naval Ammunition Depot designed and used as the world's first nuclear proving ground?**
>
> And if so, what difference does it make today?

An outlandish theory

I was doing research for a novel set in World War II when I first heard of the explosion that killed 320 Americans at the Port Chicago Naval Ammunition Depot near San Francisco, California on July 17, 1944.

Two ships were destroyed. The *SS E.A. Bryan* was obliterated; few pieces were found. Only 60 bodies were identifiable. A local newspaper reported that most victims had been "atomized."

*But **nuked**?*

Top center: two halves of the SS *Quinault Victory* sit in the Suisun Bay.

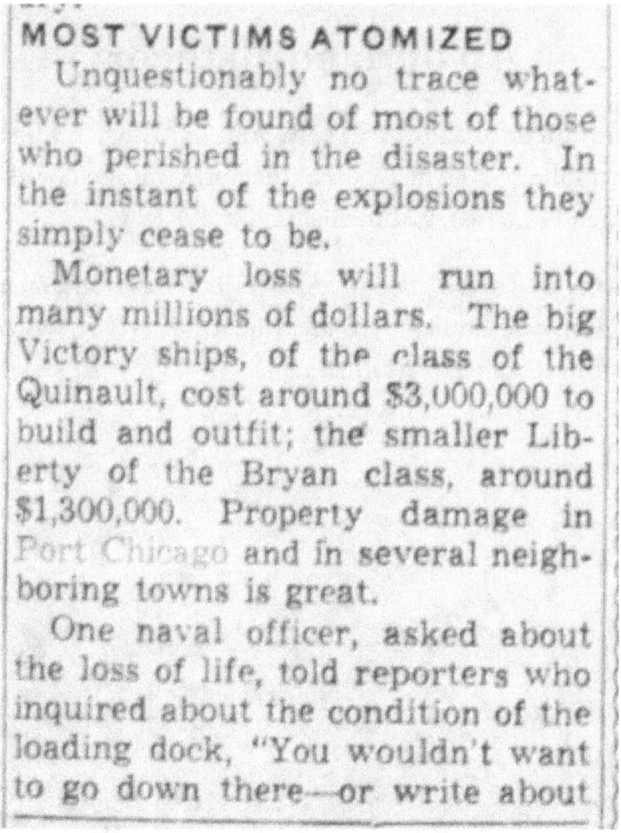

MOST VICTIMS ATOMIZED

Unquestionably no trace whatever will be found of most of those who perished in the disaster. In the instant of the explosions they simply cease to be.

Monetary loss will run into many millions of dollars. The big Victory ships, of the class of the Quinault, cost around $3,000,000 to build and outfit; the smaller Liberty of the Bryan class, around $1,300,000. Property damage in Port Chicago and in several neighboring towns is great.

One naval officer, asked about the loss of life, told reporters who inquired about the condition of the loading dock, "You wouldn't want to go down there—or write about

"*Most Victims Atomized*", **San Bernardino Sun**, Volume 50, 19 July 1944 p 2

When I began the research that led to this book, I just wanted to know if the theory made any sense whatsoever. Sure, the massive explosion was unprecedented; and yes, the Naval Court of Inquiry that investigated the explosion said the cause could not be determined. But why would anyone believe it was an intentional detonation, conducted by the Manhattan Project in the 'race to build the bomb'?

I looked around for people who could help me understand the incredible conspiracy theory. Some of the articles that supported it were obviously sensationalized, a mixture of exaggerated facts and distortion. There were a few that seemed to confirm Vogel's theory, but they did not bring much new evidence.

The few sources that opposed the theory were based on opinion and speculation, not evidence. To me, their faulty arguments strengthened Vogel's case. The only conclusion I could draw was that before I could decide one way or another, I needed more information.

Vogel's extensive work laid the foundation for a more thorough investigation. But most people who were aware of Vogel's hypothesis decided it was "just" a conspiracy theory; taboo, untouchable, not worthy of investigation.

Well, that's *almost* true. The PCnet is indeed the *unproven* **theory of a conspiracy**. And so far, other than Vogel's thirty-five-year independent investigation, it has been an ***uninvestigated*** theory. But why?

By contrast, two government agencies researched the so-called "Philadelphia Experiment". Both concluded that it was scientifically unlikely that a ship had disappeared from the Philadelphia Navy Yard in the fall of 1943.

It was possible to conduct a secret nuclear test at Port Chicago. There is nothing inherently implausible about the PCnet. It was logistically possible for the U.S. to conduct a nuclear experiment at a naval base in 1944 and to keep it a secret. After all, they built the first atomic bomb in utter secrecy.

Incredible? Yes. It's hard to believe that the U.S. government would have conducted a nuclear experiment on American soil, using American citizens as subjects. That is, we all want to believe our government could not have done such a thing. Or conducted the Tuskegee syphilis experiment; or used American Vets as guinea pigs; or conducted human radiation experiments . . .

Today, America seems to be unraveling; and we wonder why. But, as with our individual lives, where we are today is a direct result of where we have been and what we have done in the past.

I had no idea my research would be so far-ranging. I am astonished at my findings. I learned many facts related to the bomb and to World War II that have not made it into our history books. Small wonder Americans know so little about the history of the bomb; the story was designed by the bomb-makers, to sell us on the bomb.

The true history of the bomb is of great importance today because it still affects us in several ways.

For one thing, we spend untold billions on weapons of mass destruction everyone hopes will never be used; all the while fearing that they could be – at any moment; without warning.

We seldom talk about the great harm the bomb has already caused[17]. In addition to the Japanese civilians who were victims of the 1945 bombing, hundreds of thousands of people—Americans as well as citizens of other nations—struggle with severe illnesses resulting from the multi-generational effects of radioactive fallout and mismanagement of nuclear wastes.

Two other aspects of our current circumstances trace back to the bomb in a less obvious way. For one thing, there is what Robert Jay Lifton and Greg Mitchell defined in *Hiroshima in America* as "psychic numbness."[18] As a nation, we celebrated the day the bomb was dropped. After all, it was the end of the war, and we won it, with the bomb God gave us. It was a great scientific achievement, a model government program.

But today, to understand some of the social unrest in America and around the world today, we can see our nation reflected in that one grand moment when our science and technology were so magnificent, and our racial attitudes so ingrained, that the lives of 100,000 Japanese people did not matter.

To defend our right to bear arms or to support the perceived need for military supremacy, we are apt to explain that "guns don't kill, people do." The simple fact is, whether by individuals or by nations, people are killed by people with weapons. If we are to survive the twenty-first century, we must take steps to reverse the likelihood of becoming ever more sophisticated, high-tech barbarians. One of the outstanding lessons of nuclear history is the momentum that was built into the very creation of the unprecedented weapon.

Few experts still believe in the lasting power of "mutually assured deterrence" as a restraint against a first nuclear strike or against all-out war among nuclear nations. Yet that MAD policy is still the only explanation, so far, for why there has not been another war involving nuclear weapons. This precarious position suggests that if we are not to be ruled by the fact that we possess thermonuclear weapons, we must reckon with the outmoded cultural attitudes of our relatively recent history. Our social attitudes must be upgraded to *at least* match our military capabilities.

History is continuous; that's why we need to know some of the lesser-known background stories behind the bomb. For example, the people who created and promoted the "total war" policies of the 1940s were the same people who would have conducted a nuclear test at Port Chicago. The people who convinced Americans that the bomb was a good thing and its use was necessary are the same people who would have covered up a nuclear test at Port Chicago with the story of how the disaster led to integration of the Navy.

In September 1940, President Roosevelt suggested that the Navy might install Negro bands on ships as a way of getting around the demand for integration of the Navy.

In July 1942, President Roosevelt insisted that "Minorities are vital to victory." The Navy must open its doors to Negroes, he said. BuNav could "invent" something for the colored sailors to do; he had a thousand ideas of his own.

In July 1944, 202 African American sailors were among the 320 Americans killed in an explosion that may have been the proof-test of the world's first nuclear weapon.

***Mere** coincidence?*

We celebrate the Manhattan Project, but maybe we should take a second look. Regardless of whether the PCnet is accurate, the secret Manhattan Project was a political coup, a subtle and sublime attack on our democracy. We have yet to recognize the American bomb project as a secret plan, hatched by a small group of powerful men[19], to achieve a questionable objective at the expense of others. The Manhattan Project was, by definition, a grand conspiracy.

Set aside the unresolvable, opinion-based debate about the decision to use the bomb; we need to talk a great deal more about the decision to build it.

We can say what we want about whether or not it was necessary to use it; but we have yet to talk about whether or not it was necessary to create it. Not whether or not there were alternatives to using it; what were the alternatives to building it? Were they ever considered? Why, or why not?

One of the most successful American undertakings in history was the result of a conspiracy. This should be a troublesome realization.

Two important lessons I learned from my study of the PCnet:

> 1) If no one is watching, a small group of powerful people can go to extremes to get what they want; and in their quest, no one's life matters.
>
> 2) The attitudes of our leaders, and their actions, can shape not only our world but the way we see it, and how we treat one another.

It seems to me that the way to prevent this kind of thing from happening in the future is to recognize that it did (and can) happen, and then to become more conscious of what is happening now.

The more surprising information I uncovered, the more I realized that this information must be shared with the American public. I looked for people to help spread the word; but I had a dilemma: the Port Chicago nuclear explosion theory has been perceived as a "mere conspiracy theory". How was

I to show that it was important enough to be taken seriously? How could I convince anyone that the project was worthwhile?

I came to realize that it would only be worthwhile to others if it was to me. I had to take time to learn the facts and build the case. This book emerged from that effort.

The power of the past

In this book, **American Nuclear Deception**, my goal is to present the information in a way that is clear, accurate, readable and engaging. Where footnotes seemed necessary, I have tried to make them as user-friendly as possible.

There be some mistake in this book. (See what I did there?) I tried diligently to find and fix them. (The good part about mistakes, though, is that you can learn from them. That's why we look back at the past. So, if you do find a mistake, please point it out and let's get it fixed.)

The truth can only set us free when we know it. We can learn from our past mistakes, but only when we are aware of them. If the PCnet is correct, it means that, for whatever reasons they had at the time, the Roosevelt administration did something most people today would consider wrong; that is, they chose to override the moral imperative that would have prevented the test of a nuclear bomb at Port Chicago.

Who knows what might have happened if they had not crossed that first line? It may have tempered the quest to pursue the bomb. It may have given them the opportunity to consider other alternatives.

But it appears they did cross that line, and then another, and another. Today, the whole world suffers from the multiple consequences. We, Americans, must recognize this truth, and own up to it, so we can figure out, together, how to fix it.

As for the errors in this book, I invite and encourage readers to do a little fact-checking. Perhaps together we can get the story—get *our history*—straight.

Because, as the **Washington Post** slogan says, "Democracy dies in darkness". The PCnet is like a checkup, focusing on one moment in our history, to see how it contributed to the mess we're in today.

Evidence

There are many pieces of the puzzle that show the big picture of the PCnet. Each item in the list below is a major clue in itself. They will each be discussed at length in the following chapters, along with other elements that are less prominent.

In addition to the significance of each item, several factors hold these clues together:

1. There are so many of these unknown facts, and they appear to have been hidden in history.
2. Most have enough historic significance in and of themselves to have been widely known.
3. Even where there are individual explanations for why they are unknown, their overlap makes it even unlikely that their omission from history is a simple oversight.
4. Individually and combined, these facts all support the PCnet.
5. In some cases, the PCnet is the best explanation for what happened and why it is unknown.

Evidence in support of the PCnet is found in a wide range of topics:

The early history of the atomic bomb is full of gaps and unanswered questions. The original, misleading history was deliberately concocted by government officials and Manhattan Project leaders to sell the bomb to the American public and avoid Congressional hearings on the secret project. Many unknown secrets of nuclear history come to light in an investigation of the PCnet.

The famous Einstein letter story is one of the false premises that historians have reconsidered. Einstein did not write the letter himself; and the letter had little influence on FDR's decision to create the bomb.

Alexander Sachs, FDR's economic adviser, was the first person to introduce him to the bomb; but in the traditional history, he appears to be little more than the courier who hand-delivered the so-called Einstein letter. Sachs was sometimes the only man in the room; yet historians have downplayed or overlooked this unique source of certain information about FDR and the bomb.

President Franklin D. Roosevelt was the president who unilaterally authorized the American bomb project, yet historians have only recently begun to scrutinize his decision to pursue it and his plans to use it. Since the compelling need to create the bomb drove the need to test it, FDR's motives provide a major key to the PCnet.

President Harry Truman is the president most associated with the bomb, though the decision to use it was inherent in the decision to create it. Historians insist that Truman knew nothing about the bomb until after his inauguration, despite clear evidence that he did. How did the "accidental president" come to preside over one of the major decisions of the 21st century? The upheaval that led to his nomination at the Democratic National Convention in July 1944 may have implications for the PCnet.

Vannevar Bush, head of the Office of Scientific Research and Development (OSRD), reported directly to FDR. Bush coordinated the research that led to the Manhattan Project. Bush recommended his Special Assistant, Captain William S. Parsons, USN, to head ordnance for the Manhattan Project. Bush administered both the Manhattan Project and the Committee on Medical Research (CMR), which was the agency that sponsored the Tuskegee syphilis study. From the beginning, Bush was aware of the eventual need for a remote location to test the weapon.

William S. Parsons is best known as the Navy captain who armed the *Little Boy* bomb in midair as the *Enola Gay* flew to Hiroshima; but he was an expert Experimental Officer, instrumental in the development of each of the three key military technologies of World War II—radar, the proximity fuze, and the atomic bomb.

As head of the Ordnance division of the Manhattan Project, Parsons was the man responsible for the production and delivery of the bomb. Parsons, the man most likely to sponsor a clandestine test at Port Chicago, visited the site of the explosion in person, to study "the effects of the detonation." Parsons made many statements that support the theory of a nuclear test at Port Chicago. The secret PCnet may be the reason why the first "atomic admiral" is such an obscure historical figure.

The Navy was allegedly left out of the bomb project, but the service made several contributions to the project, including the participation of key personnel and transportation of the bomb to the Pacific. Omission of the Navy's role in the Project helps to conceal several clues to the PCnet.

Albert Einstein worked on torpedo design as a consultant for the Navy during the time Parsons was pursuing the idea of a nuclear torpedo. One of Einstein's Navy liaisons, Lt. Stephen Brunauer, went on to serve as Officer-in-Charge at *Operation Crossroads*, the 1946 tests of the atomic bomb against Naval ships; Parsons served as Deputy for Technical Direction.

The *Thin Man* plutonium gun-assembly bomb was the first prototype of an atomic bomb; but it gets short shrift in history. The possibility that plutonium would be ineffective in a gun-assembly bomb was discussed as early as April 1943, but the work on the *Thin Man* continued until it was "shelved" on July 17th, 1944. the day of the Port Chicago explosion. The mystery of the disappearing *Thin Man* helps explain how the *Little Boy* uranium gun bomb could be used at Hiroshima without ever being tested.

FDR's trip to the West Coast in July 1944—during the same week of the DNC in Chicago, Illinois and the massive explosion near San Francisco—was one of several historical coincidences that are best explained by the Port Chicago nuclear explosion theory.

The *Thin Man* bomb was "shelved" on July 17th, 1944, marking a turning point in the Manhattan Project. The little-known history of the *Thin Man* holds clues to the secret history of the atomic bomb project.

The Port Chicago explosion provided essential data for the research and development of the bomb. Clues to the PCnet are found in the link between the Project and the explosion, and in the overlapping timelines of the Manhattan Project and the Port Chicago Naval Ammunition Depot.

The Port Chicago Naval Ammunition Depot was "an accident waiting to happen." The remote military facility was a controlled site where Manhattan Project engineers could camouflage the full-scale test that provided useful, real-world data on the effects of an atomic detonation.

The Port Chicago court martial quickly overshadowed the historic explosion. The Navy went to great lengths to publicize the mass tribunal, a sleight of narrative-weaving that transformed the story of the massive explosion into the story of the mass mutiny trial. A major event in American history was relegated to the pages of "African-American" history.

The Port Chicago nuclear explosion theory is a serious and extreme allegation; it requires substantial proof. As we will see, there is a great deal of wide-ranging evidence to support it.

Note:

American Nuclear Deception presents evidence in support of the Port Chicago nuclear explosion theory (PCnet). Although some elements are more compelling than others, no single piece of evidence forms a "smoking gun" that seals the case.

On the other hand, taken together and read in context, the full body of little-known evidence suggests a strong likelihood that the Port Chicago explosion was indeed the result of a secret test conducted in connection with the Manhattan Project.

***For the sake of discussion*, I argue the case from the *assumption* that the theory is true.** Certain facts are repeated throughout this book, first because it requires this full context to appreciate their significance, but in many cases because the little-known facts are significant enough to bear repetition.

The PCnet has <u>not</u> been proven.

Nor has the theory been debunked.

This book, like a grand jury hearing, is designed to explore the evidence, to see if there is a valid case to be made. That verdict must rest with the jury—with you.

Terms to know

This is essentially a glossary, but I placed it up front to help provide a sense of the setting as you read the rest of the book. Pay closest attention to the terms that are starred. They are especially important to the PCnet.

air blast – one of four deadly effects of a nuclear explosion, including radioactivity, shock wave, and nuclear heat

ammunition dump* - a military base where ammunition is produced or stored, or both; another name for Naval magazine

Army Air Force – the Air Force was part of the Army during World War II.

atomic bomb – The first weapons of mass destruction produced by the Manhattan Project, by hand, were called "atomic bombs". They were toys compared to today's thermonuclear weapons. (This book discusses the history of atomic bombs up till July 1946.) See also *Thin Man, Little Boy, the Gadget, Fat Man*.

atomic diplomacy - use of atomic bomb to coerce other nations to cooperate; nuclear history "revisionists" believe President Truman used the bomb to manipulate Russian president, Joseph Stalin; see *big stick*

Atomic Vets – American servicemen who were used as guinea pigs in nuclear tests; they were told what to expect; that it was a beautiful sight; and that they were not to tell.

big stick* - use of power status by one nation to coerce others to cooperate; see *atomic diplomacy*

chain reaction - a series of nuclear fissions; if unchecked, the power builds up until it becomes a nuclear explosion (also known as nuclear chain reaction)

collywobbles* – On July 20th, 1944, hours before he gave his acceptance speech for a fourth term, FDR had a severe attack of pain that left him writhing on the floor of the presidential train. He later dismissed the attack, calling it the "collywobbles."

COMINCH* – Commander-in-Chief; The President of the United States is the COMINCH over all the U.S. armed services

compartmentalization – the (military) system used to reduce flow of information only to those who have a "need to know" in order to complete their particular assignment; prevents people from putting information together to see the big picture.

conspiracy* – a secret plan, devised by a small group of people who have the power to carry it out, usually to achieve a questionable objective, at the expense of others; see *theory of a conspiracy*

conspiracy theory* - *unproven allegation* that there has been a conspiracy; generally used with air quotes, or to otherwise signal that the conspiracy theorist is kooky; ***see** theory of a conspiracy*

conspiracy theorist* - a person who promotes conspiracy theory; especially one who sees "conspiracies everywhere"; see *theory of a conspiracy*

court martial* – a military trial; legal code differs from civilian law; military trials are generally closed to the public

colored people - an acceptable alternative to Negro in the 1940s; see *people of color*

detonation* – the act of triggering an explosive device, including some bombs; see *predetonation, spontaneous fission, fizzle*

double victory - African Americans (then called Negroes) fought for victory at home, including equal justice and jobs, as well as victory overseas

Downwinders – citizens of the Midwest (Arizona, New Mexico, Nevada and Utah) who were exposed to radioactive fallout from nuclear tests. see *fallout*

Einstein letter* – the famous letter *supposedly* written by Albert Einstein to warn President Roosevelt about the potential of a German atomic bomb; in fact, Einstein signed the letter but did not write it; and its purpose was to gain government support for uranium research. See *race for the bomb*

fallout – the spread of radioactive particles after a nuclear explosion

Fat Man* Dummy – one of two bomb models in early tests; used to train pilots and test bomb mechanism (also known as Pumpkins)

fizzle* – a failed nuclear explosion; one that fails to reach its minimum effect; see *predetonation, spontaneous fission*

flagship* – the lead ship in a fleet

fleet* – convoy; ships that are part of a convoy on a particular mission

FOIA - Freedom of Information Act; law allowing citizens to request to have specific documents declassified

fourth term – President Franklin D. Roosevelt was the only U.S. President to serve four terms. He was four months into his fourth when he died on April 12, 1945. After his death, Congress passed the law that limits Presidents to two terms.

Freedom of Information Act (FOIA) – law allowing citizens to request to have specific documents declassified

Gadget – code name for the first implosion (atomic) bomb tested at Alamogordo, New Mexico on July 16, 1945 at the Trinity test, the world' first (known) nuclear explosion.

Hiroshima – Japan city where the first atomic bomb killed approximately 100,000, mostly civilians, on August 6, 1945

Honolulu Conference* – FDR's meeting with his Pacific Commanders, General Douglas MacArthur and Admiral Chester Nimitz

Human Radiation Experiments (HRE) – experimentation on unaware American citizens to study plutonium

Little Boy* – the uranium gun-assembly bomb; successor to the *Thin Man*. The U.S. dropped the *Little Boy* on Hiroshima, Japan, on August 6, 1945. It was the first of two atomic bombs ever used in an attack against another nation.

Los Alamos* – the New Mexico laboratory of the Manhattan Project, where the scientists and engineers produced the actual bomb

Manhattan Project* – code name for the U.S. project that built the world's first atomic bomb during World War II

Mare Island* – the Mare Island Naval Shipyard, near San Francisco, California; across the Bay from Port Chicago, a 35-mile drive by land

Marshall Islands* – a group of islands in the South Pacific, including the Bikini, Eniwetok and Rongelap atolls. After the war, the (newly created) United Nations gave the U.S. trusteeship over the Marshall Islands. In July 1946, the U.S. conducted *Operation Crossroads*, the test of atomic bombs against Naval ships. It was the first of 67 nuclear tests in the islands between 1946 and 1958.

mushroom cloud – a mushroom shaped cloud filled with debris from a nuclear explosion (allegedly possible in non-nuclear explosions as well)

mutually assured destruction - MAD; nuclear deterrence policy; presumably nuclear nations restrain from war, and in particular resist a first nuclear strike, because the certainty of powerful immediate retaliation means no one can win a nuclear war. According to some experts, the policy is an untested theory that may be outmoded, especially given the possibility of "bad actors" who are not bound by the unofficial restraint.

Nagasaki - Japanese city where the second atomic bomb killed estimated 80,000, mostly civilians, on August 9, 1945

naval magazine* – a military base where ammunition is produced or stored, or both. The PCNAD also shipped ammunition overseas. also called ammunition dump

ordnance* - weapons, and other war material

PCNAD* - Port Chicago Naval Ammunition Depot (PCNAD); naval magazine, ammunition dump

Nevada Proving Grounds – the first nuclear proving grounds on the American continent

nuclear chain reaction – a series of nuclear fissions; if unchecked, the power builds up until it becomes a nuclear explosion (also known as chain reaction)

nuclear fission – the energy that is released when the atom is split

nuclear heat – one of four deadly effects of a nuclear explosion, including radioactivity, air blast, and shock wave

nuclear torpedo* - a torpedo designed to destroy a target by underwater delivery of an atomic weapon; one of the early ideas pursued by the ordnance division of the Manhattan Project, lead by Capt. William S. Parsons. As stated in the Szilard-Einstein letter, scientists originally thought an atomic bomb would be too heavy for an airplane.

Operation Crossroads* - the 1946 test of atomic bombs against Naval ships, on a grand scale. See Test ABLE and Test BAKER; the first of 67 nuclear tests conducted by the U.S. between 1946 and 1958

people of color - non-white people throughout the world: includes Africans, African Americans; Asians, Caribbean, South Americans and others; see *racism, colored people, Negroes*

predetonation* – the detonation of a bomb before an explosion reaches its maximum potential; see *spontaneous fission, fizzle*

plutonium* – a man-made element derived from uranium

POCH - The Port Chicago National Memorial (POCH) is a National Park unit at the site of the July 17, 1944 explosion. The city of Port Chicago no longer exists; the Navy bought the city in 1968 and razed it.

Port Chicago* - Port Chicago generally refers to the Naval Ammunition Depot (PCNAD) on the West Coast, near San Francisco, California. The base got its name from the nearby town, Port Chicago, which was razed by the Navy in 1968. (Port Chicago is not the same thing as Port *of* Chicago and has *nothing to do* with the city in Illinois.)

Port Chicago court martial* - Fifty of the sailors who survived the explosion refused to load ammunition onto ships, citing fear of another explosion. The 50 sailors were each found guilty of conspiracy to commit mutiny. Their mass court martial was the largest trial in U.S. Naval history. (Also called Port Chicago mutiny trial)

Port Chicago disaster – The disaster consisted of the "Port Chicago explosion" and the "Port Chicago mutiny trial" (also referred to as the "Port Chicago court martial")

Port Chicago explosion* – A massive explosion occurred at the PCNAD on July 17, 1944. Two ships were destroyed; 320 people killed and at least 390 seriously injured.

Port Chicago 50 – the fifty black sailors found guilty of conspiracy to commit mutiny; their sentences were reduced after the war and most were restored to service. One of them (Freddie Meeks) received a pardon from President Clinton in 1999. Each year, Congress considers new legislation to exonerate the Port Chicago 50.

Port Chicago mutiny (trial)* – Fifty of the sailors who survived the explosion refused to load ammunition onto ships, citing fear of another explosion. In a mass court martial that was the largest trial in U.S. Naval history, each of the 50 sailors was found guilty of conspiracy to commit mutiny. The sailors became known as "The Port Chicago 50".

proximity fuze – one of three key technologies developed during the war; the other two were radar and the atomic bomb. (William S. Parsons, then a Captain in the Navy, was instrumental in the development of each. As head of ordnance for the Manhattan Project, he was responsible for producing the bomb.)

pumpkins - bomb models used to train pilots and test bomb mechanism

race for the bomb* - the traditional rationale for America's bomb project was that America began its own project after FDR in response to a warning from scientists who were afraid Hitler was working on an atomic bomb. But the real race to get the bomb first was based on the common knowledge that the first nation to get the bomb would have a distinct military advantage. Germany did not actively pursue an atomic weapon during the war; for several reasons, including other priorities. The U.S. had unique advantages, including: geographical space for the necessary factories; manpower; economic power; geographical distance from the war; a concentration of eminent scientists, including emigrants from Europe; a strong industrial base. And, of course, the flexibility of democracy under a powerful, visionary and independent leader with global ambition.

racism - ***spectrum of bias*** against people of another race. Expressions of racism range from individual attitudes (prejudice, hatred) . . .to collective social attitudes . . . to legal discrimination . . . to systemic racism. Powerful people sometimes exploit the racism of social groups to achieve other political or social ends.

radioactivity – the emission of particles blast from a nuclear explosion; one of four deadly effects of a nuclear explosion, including shock wave, air blast, and heat

secrecy* – the set of policies and procedures used to guard nuclear information. Several strategies were used to keep the Manhattan Project secret ***from the American public***. Examples of deliberate deception revealed through the lens of the PCnet include:

- administrative confusion: overlapping assignments; backbiting, spreading rumors

- black budget: managing financial accounts so as to hide questionable allocations

- camouflage: using ordinary setting to veil dubious activity

- censorship: bar on publication of specific information; see *voluntary self-censorship*

- circuitous communication: routing messages and deliveries through other parties or offices

- classification: labeling documents - confidential, secret, top secret, eyes only

- coded language: code names, coded messages

- compartmentalization: separation of information into bits, like puzzle pieces, to hide big picture

- distractions (also known as 'red herrings'): creation of one spectacle to detract attention from another (as in the case of the Port Chicago mutiny trial)

- false information: deliberate lies

- misinformation: manipulation of facts; changing minor elements of information to appear valid

- "mission orders": verbal orders used to keep from creating a paper trail

- omission of key information: omission or deletion of names, facts or other details ordinarily expected

- secret files; also known as 'safe files': hiding documents in unexpected places

- subterfuge, intrigue: spy stuff – secret rendezvous, disguises, false identity, and other methods

- vague language: lack of response to questions; unexplained riddles; puzzling statements

secret cities – the three cities that were created for the Manhattan Project: Hanover, Washington, where plutonium was manufactured; Oak Ridge, Tennessee, site of the uranium enrichment plant; and Los Alamos, New Mexico, where the bomb was built. Employees could not communicate with their families and were not allowed to leave the compounds, which were fenced in and guarded, for the duration of the war.

shock wave - one of four deadly effects of a nuclear explosion, including radioactivity, air blast, and heat

spontaneous fission* - chemical reaction resulting in unplanned detonation; see *predetonation, fizzle*

SS *E. A. Bryan** – one of two ships demolished at Port Chicago; the Bryan was obliterated; few pieces remained.

SS *Quinault Victory** - one of two ships demolished at Port Chicago; the Quinault Victory was left standing in the Bay, broken in half

supremacy* - the idea that one (group, race, nation, worldview . . .) is superior to all others, perhaps by divine right

Thin Man* – the world's first atomic bomb model, a plutonium gun-assembly bomb. Due to a predetonation problem, the *Thin Man* never went into production. The *Little Boy*, its successor, was sometimes called the *Thin Man. (The Thin Man bomb gets very short shrift in the history of the bomb. The PCnet may explain why.)*

***Thin Man* Dummy** – one of two bomb models in early tests; used to train pilots and test bomb mechanism (also known as "pumpkins")

Thirteenth Naval District – The Naval jurisdiction on the West Coast that includes (among other locations) San Diego

torpedo* – underwater bomb; The torpedo, a new technology in World War I, was still under development in World War II.

Trinity* – code name for the world's first (known) nuclear explosion when the *Gadget* was detonated at Alamogordo, New Mexico on July 16th, 1945. No one knows why J. Robert Oppenheimer called the test "Trinity." Many assume it relates to a poem by John Donne. Oppenheimer said otherwise but still could not explain.

(*A little-known pre-test conducted in May is referred to as "the past Trinity" in the minutes from a meeting preparing for the Trinity test. This is of interest to the PCnet because it <u>may</u> have been the <u>second of three tests</u>: in July 1944, May 1945 and July 1945*

torpedo scandal – failure of American torpedoes between 1941 and 1943 due to lack of adequate testing

Truman Committee* – Congressional committee established in 1941 by Senator Harry Truman. The powerful committee had oversight over defense spending.

Twelfth Naval District* – The Naval jurisdiction on the West Coast that includes (among other locations) the Mare Island Navy Shipyard and the Port Chicago Naval Ammunition Depot (later named Concord Naval Weapons Station (CNWS); more recently named Military Ocean Terminal Concord (MOTCO))

underwater delivery* – the method of bombing a target with an underwater weapon; a torpedo or depth charge. As the initial research leading to the bomb got underway in 1939, the thought was that it would be too large to carry in an airplane. (*see* nuclear torpedo)

uranium – the element used when scientists discovered that a tremendous amount of energy is released when an atom is split. The process is called nuclear fission. (Nuclear fusion is a different process that also creates a nuclear explosion. Today's thermonuclear weapons use fusion; both topics are beyond the scope of this book, which focuses on the early history of the atomic bomb.)

USS *Baltimore* (CA-68)* – the heavy cruiser that was COMINCH flagship for the FDR's cruise to Hawaii for the Honolulu Conference in July 1944

voluntary self-censorship*: self-policing by the press to honor government bar on publication of specific information (reinforced by the U.S. Office of Censorship)

PCnet people

Three atomic musketeers

The roles of three key people were downplayed, omitted and hidden in the traditional history of the bomb, but these three men had direct influence on the creation and use of the bomb.

PRESIDENT FRANKLIN D. ROOSEVELT wanted America to be the first nation to get the bomb. He had ambitions for the United Nations, and even considered resigning from his fourth term to head the new world-governing organization.

VANNEVAR BUSH, Office of Scientific Research and Development[20] recognized the importance of military science. As war began to rumble in Europe, he moved to Washington to get a "piece of the action," positioning himself to run all of the nation's science projects during the war as FDR's tsar of science. He administered the Manhattan Project and the Committee on Medical Research (the agency that conducted the Tuskegee syphilis experiment.)

WILLIAM S. PARSONS, who headed the ordnance division for the Manhattan Project, was working as an Experimental Officer at the Dahlgren Proving Ground in Virginia in 1939 when he became Special Assistant to Vannevar Bush. His successful leadership in developing the new proximity fuze (one of three key military technologies of WWII) prompted Bush to recommend the exemplary engineer to head the ordnance division of the Manhattan Project. Though he was responsible for the production and delivery of the bomb, Parsons is best known to history as "the Navy captain who armed the *Little Boy* bomb as the Enola Gay B-29 bomber carried its fateful cargo to Hiroshima."

Together, this powerful triad had the means, money and motives to conduct a nuclear test at Port Chicago. In addition to their individual contributions to the bomb, their little-known connection is a major clue to the PCnet.

They had a need to build the bomb: there would be no better opportunity than during a war that began just as scientists discovered nuclear fission.

And if they had a need to test the unprecedented weapon, there could be no better location than the one they would create for that purpose.

Three of the key people most commonly associated with the bomb in historical accounts are President Harry Truman, General Leslie R. Groves, and physicist J. Robert Oppenheimer.

Harry S. Truman is the U.S. President whose name is most often associated with the bomb. That's because, although FDR presided over its creation, Truman presided over the decision to use it against Japan in 1945.

In 1941, Senator Truman sponsored and headed a Congressional committee that had oversight of defense spending. But Secretary of War Henry Stimson and General George Marshall each directed him to stay clear of an investigation into the huge, mysterious construction site at Hanover, Washington, and Truman complied.

Three years later, Truman allegedly resisted but ultimately accepted the charge to become Roosevelt's running mate when the popular wartime president was nominated for his fourth term. Truman's upset victory over Vice-President Henry Wallace came about amidst the raucous and manipulated Democratic National Convention of mid-July 1944, which took place during the same week of the Port Chicago explosion, while FDR was on the West Coast.

When FDR died on April 12, 1945, Truman became the "accidental" president—and "stepfather" of the bomb. Most historians insist that Truman did not know about the bomb until after his inauguration. (**The Smyth**

Report – the first official history of the atomic bomb – seems to say otherwise.)

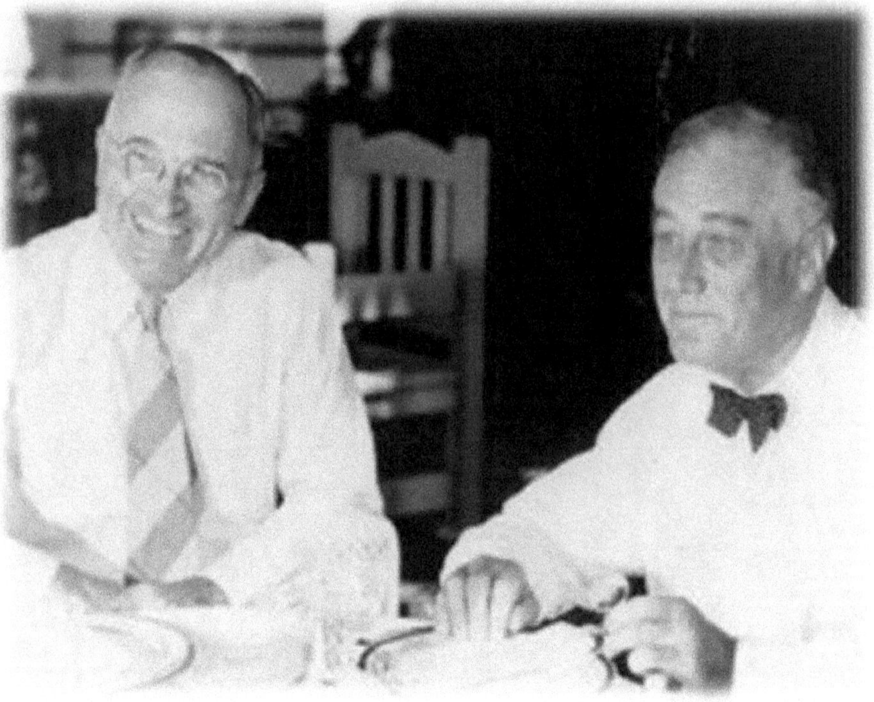

Running mates: (l) **Senator Harry S. Truman**, Vice Presidential Candidate at lunch with **President Franklin D. Roosevelt** (FDR) on the White House lawn - August 18, 1944

Leslie R. Groves was Director of the Manhattan Project. Groves, a U.S. Army Corps of Engineers officer, was promoted to Lt. General and assigned to the Manhattan Project after successfully overseeing construction of the Pentagon. By completing that project ahead of schedule and under budget, the tall, burly and blustery, no-nonsense officer established his reputation for getting the job done. Groves enforced a system of compartmentalization (sharing information on strict need-to-know basis) as a way of ensuring secrecy.

J. Robert Oppenheimer, the brilliant director of the Los Alamos laboratory, was skinny, witty and sharp-tongued. Groves had to intervene to get Oppenheimer's security clearance because he had been a Communist sympathizer (and possibly because he displayed some emotional instability as a youth).

Groves and Oppenheimer pose for postwar publicity shots, including at Trinity test site.

Although historical accounts say the Navy was omitted from the bomb project, Parsons was one of several naval officers who played key roles in the early history of the bomb.

James R. Crenshaw was one of three men on the Naval Court of Inquiry (COI) that investigated the Port Chicago explosion. The COI concluded that the cause could never be determined. Capt. Crenshaw was the long-time friend and brother-in-law of Capt. William S. Parsons, head of ordnance for the Manhattan Project.

Stephen Brunauer, a Lieutenant in the Navy during World War II, served as liaison between Einstein and the Navy Bureau of Ordnance. Brunauer and von Neumann visited Einstein together on at least one occasion.[21] In the summer of 1946, Brunauer was promoted to Commodore and served as Officer-In-Charge at *Operation Crossroads*, for which Parsons was Technical Director.

Official U.S. Navy photo

REAR ADMIRAL C. H. WRIGHT, U.S.N., Commandant, 12th Naval District and Naval Operating Base.

Rear Admiral Carleton H. Wright was awarded the Navy Cross in 1942, despite losing several ships in a battle against Japan. He was transferred to shore duty in 1943. In 1944, he became Commander of the Twelfth Naval District. Wright, who had a brief meeting with Parsons on July 20th, had already authorized the Port Chicago Court martial on July 14th, three days before the Port Chicago explosion.

After the explosion, Wright praised several Negro sailors who acted with especial valor in the immediate aftermath of the explosion, saying their conduct was "as was to be expected." Weeks later, however, when 328 sailors cited fear of another explosion as the reason why they refused to load ammunition onto ships, Wright called them cowards. He said he would see to it that they got the death penalty for mutiny in time of war.

Three collaborators played key roles in the creation of the bomb; their relationship is reviewed in the PCnet.

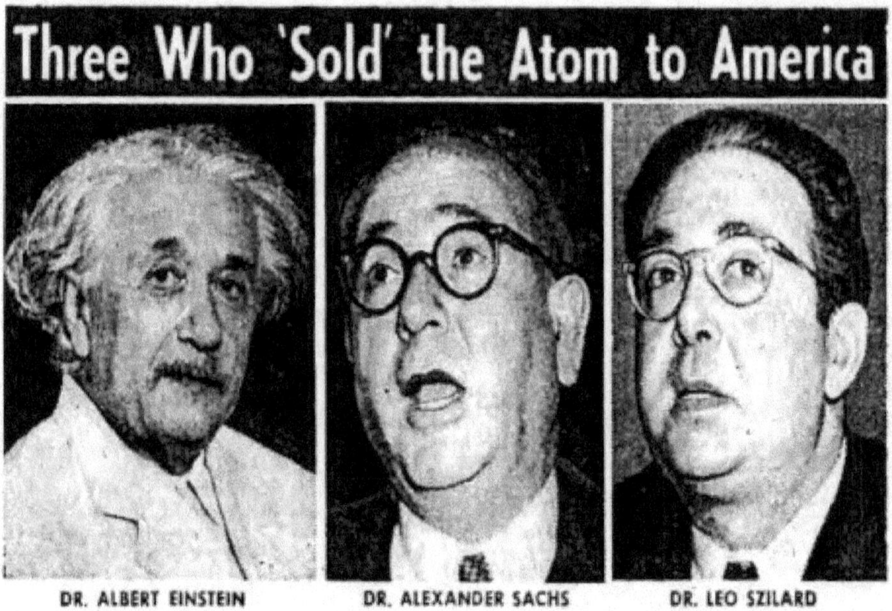

"Three who 'Sold' the Atom to America," in "The Atomic Future" - Howard W. Blakeslee, Associated Press supplement, May 1946, from the Alexander Sachs papers p 125[22].

Albert Einstein, the beloved genius, *signed* the Einstein letter that supposedly led FDR to pursue the bomb. (As historians now acknowledge, the famous letter was *not* the catalyst for the U.S. atomic bomb project.) Although he later said signing the letter was his only contribution to the U.S. bomb project, Einstein endorsed the misleading story by posing for a picture with fellow physicist Leo Szilard, now recognized as the real author of the letter. Whether with or without his knowledge, **Einstein's work on torpedo design for the Navy** may have supported Parson's quest for a nuclear torpedo.[23]

Leo Szilard was one of a small group of emigrant scientists, including Eugene Wigner, Edward Teller and Enrico Fermi, who realized the potential danger of an atomic bomb, especially if Hitler got it first. When Sachs heard of their attempts to gain government support for uranium research, he suggested writing a letter to the President. Einstein did not conceive of the letter, and he did not write it himself. He did participate in its creation, though, and his signature was intended to give it more clout.

Alexander Sachs is best known as the man who was asked to deliver the "Szilard-Einstein" letter (as it is now called) to the White House, because, as FDR's long-time friend and advisor, he had access to the President. In fact, the Wall Street economist had already spoken with FDR about the bomb in the spring of 1939. That summer, he worked with the scientists to produce the famous letter, which was part of a dossier he used in support of his extended presentation to FDR on October 11th and 12th. Two weeks later, FDR set up the Uranium Advisory Committee (UC). Sachs continued to serve as FDR's liaison, representing the President's interest to scientists and others. He helped administer the UC; facilitated the import of uranium from the Belgian Congo; coordinated with the scientists, and contributed to the Project in other ways.

The contributions and connections of these key people are not altogether evident in the traditional accounts, but they become both evident and significant when the history of the bomb is reviewed through the lens of the PCnet.

The story of the bomb story

The history of the bomb typically begins with the legendary story of Albert Einstein writing a letter to President Roosevelt to warn him that Germany might be working on an atomic bomb. In the original version, Einstein himself wrote the letter and had it hand-delivered by Alexander Sachs, a friend and advisor to FDR who had access to the White House.

In the updated version, historians acknowledge that Einstein signed the letter at the request of his friend and colleague, Leo Szilard, a Hungarian emigrant who was the first person to figure out how a nuclear bomb might work.

Szilard and Einstein, along with several other refugee scientists, tried to alert the President after they failed to get support from other government agencies.

Sachs, the economist, was not just a courier for Einstein and Szilard, as the traditional narrative suggests. Several historic records say he was the man who first convinced FDR to pursue the bomb project.

As we will see, there are other aspects to the story, as well as at least two other versions.

Upon closer scrutiny, it becomes apparent that the famous letter was not the catalyst for the bomb project after all. The letter has been interpreted as a warning about the threat of a German bomb, but the scientists were actually seeking government support for their promising uranium research. Other aspects of the letter provide more direct evidence for the PCnet.

Sachs was much more than a messenger. In October 1939, when he carried the dossier containing the Szilard-Einstein letter to a White House meeting with FDR, Sachs was a financier, a Vice-President at Lehman Brothers. Their meeting began on the 11th and resumed the next morning. As noted, the letter was one document in a more complete dossier. There is no evidence that the President ever read it, since Sachs made an extensive and persuasive presentation, convincing the President that the atomic bomb would be an "exponential weapon." FDR agreed that pursuit of a nuclear bomb might be a "providential challenge and opportunity for Americans and democracy."

According to some accounts, a mutual friend put Sachs in contact with Szilard, a physicist at Columbia University who emigrated to the U.S. in 1938 and conducted research on uranium. (Szilard's association with Einstein dated back as far as 1930, when the two co-designed a wireless refrigerator.)

In January 1939, Niels Bohr, the distinguished physicist from Denmark, had announced the discovery by German scientists that an unprecedented burst of energy resulted when an atom was split. This discovery, called "nuclear fission," stunned the scientific community.

Thanks in part to the 1914 science fiction book by H. G. Wells, *The World Set Free,* which depicts the horrors of atomic war as a way to end war for all time, the discovery of nuclear fission was also of interest to economists, politicians and ordinary curious people.

One of the people interested in Bohr's announcement was Alexander Sachs, who had been following developments in atomic research. He reported the news to FDR. That summer, Sachs collaborated on the famous letter. Sachs claimed that he was the person who persuaded Leo Szilard to send it to FDR under Einstein's signature. He later came to resent the popular and misleading simplification of the true history.

In his biography, **A Piece of the Action**, Vannevar Bush, FDR's tsar of science during the war, gives another account of the bomb's origin. He explains why even the greatest scientists did not have such strong influence on U.S. foreign policy, war policy and military spending. By contrast, Alexander Sachs, a long-time economic adviser to FDR, had a history of impacting the President's decisions.

Along with other significant contributions Sachs made to the bomb project, he participated on the Advisory Committee on Uranium and helped secure uranium from the Belgian Congo. Sachs was also the first person to testify at the Senate Hearings on the bomb in November 1946.

So why is Sachs best known as the presidential advisor, a long-time friend of FDR, who was chosen by Szilard and Einstein to deliver the famous letter to the White House, simply because he had access to the President?

Eminent historian Barton J. Bernstein warns scholars that Sachs may not be a reliable witness. However, in some cases, the Sachs papers are the only primary source of information written at the time about events as they took place. This in itself makes his papers important; it also means they must be scrutinized with care.

The Port Chicago nuclear explosion theory (PCnet)

The Port Chicago Naval Ammunition Depot (PCNAD) was established on the San Francisco Bay in 1942. Its official purpose was to ship weapons to the Pacific theatre during World War II; but the massive explosion at the ammunition dump on July 17, 1944 played a vital role in the creation of the atomic bomb.

Photo # NH 96823 Damage at Port Chicago, Ca. View looking north toward pier.

Damages caused by the July 17, 1944 explosion at the Port Chicago Naval Ammunition Depot near San Francisco, California. Official photos, Mare Island Navy Yard

On July 17, 1944, a massive explosion rocked the pier, destroying two ships and killing 320 Americans, including 202 of the Black sailors who handled ammunition at the depot.

The story of the disaster soon came to be the story of "The Port Chicago Mutiny." On August 9th, 328 of the Black sailors refused to load ammunition onto ships anymore, claiming fear of another unexplained explosion. Most returned to work after learning that they could face the death penalty for mutiny in time of war. But fifty of them stood their ground; their highly publicized trial was the largest mass court martial in Naval history.

Unlike other race-related incidents during the war, there was no violence involved in the "work stoppage" that led to the highly publicized Port Chicago court martial; yet it was the largest mass trial in U.S. Naval History.

Annual commemorations to honor the victims of the disaster are held each July at the *Port Chicago Naval Memorial*, a unit of the National Park Service.

And, each year, Congress considers new legislation to exonerate the "Port Chicago 50."

Robert Allen, author of **The Port Chicago Mutiny**, addresses the audience during the commemoration at the *Port Chicago Naval Magazine National Memorial* which was dedicated in 1994.

The Port Chicago / Manhattan Project connection

Few people are aware of the important role the Port Chicago explosion played in the research and development of the atomic bomb. Manhattan Project scientists and engineers studied the effects of the massive explosion, relying on the Port Chicago data to design the unprecedented weapon.[24]

MEMORANDUM

24 July 1944

For: Rear Admiral W. R. Purnell, USN

From: Captain W. S. Parsons, USN

Subject: Port Chicago Disaster, Preliminary Data

1. On 20 July I arrived in San Francisco, accompanied by Ensign Reynolds and Dr. Shapiro. I called on the Commandant (Rear Admiral C. H. Wright) and discussed the disaster briefly with him, emphasizing that my mission was to obtain data on effect rather than the cause of the detonation. Admiral Wright had been in command of the Naval Mine Depot, Yorktown, Virginia, as recently as 1940, and his executive there had been Captain A. G. Cook, now Captain of the Yard at Mare Island. Admiral Wright had, therefore, considered Cook well qualified and had appointed him President of the Court of Inquiry, with Captains J. S. Crenshaw and W. B. Holden as members.

The cooperative spirit and helpfulness shown by Admiral Wright and his office was later found to obtain throughout the organization, including the Navy Yard, N.A.D. Mare Island and at Port Chicago. There was not the slightest tendency to cover up or stand on dignity or to raise questions regarding cognizance or infringement on the duties of the Court.

2. My party arrived at Mare Island about noon 20 July and, with Captain Crenshaw, proceeded to Port Chicago. Enroute we [redacted]

Various eyewitnesses interviewed by ONI agreed that the column of fire rose vertically and then mushroomed. A second detonation is supposed to have followed the first one at an interval estimated from 1/5 second to many seconds.

Comparing loss of life to the Halifax disaster, it appears that all but some five of the victims at Port Chicago were right on top of the explosion, in a position corresponding to some 25 crew members and fire fighters at Halifax. Thus, the comparison for remote victims is Halifax about 1,475, Port Chicago less than 5. If the two explosions are considered to be of the same order of magnitude, the difference in loss of life can be attributed to the fact that Port Chicago was designed for large explosions.

5. It is emphasized that the data and discussion given in this memorandum are preliminary and necessarily rough. Several weeks will be required to assemble and analyze data for an adequate report.

W. S. Parsons
Captain, USN

WSP/hg

"Memorandum, Port Chicago Disaster, Preliminary Data, 24 July 1944". (This appears to be an internal Navy memo, but Parsons was head of ordnance for the Manhattan Project and Purnell represented the Navy on the Military Policy Group.) [underlining added]

Captain William S. Parsons, head of ordnance for the Los Alamos Laboratory in Alamogordo, New Mexico, was the Naval officer in charge of producing, testing and delivering the bomb. He and his team visited the scene of the blast to study "the effects of the detonation."

In his preliminary report, Parsons noted that "Port Chicago was designed for large explosions."

The Port Chicago nuclear explosion theory supports a literal interpretation of Parsons' statement. Wide-ranging evidence supports the unproven hypothesis that the PCNAD was designed and used as the world's first nuclear proving ground.

“In typical Port Chicago fashion”

Admittedly, the idea of a nuclear explosion at Port Chicago seems ridiculous on its surface. In the first place, the theory contradicts what we already know about the bomb. We recall, for example, that the famous Trinity test conducted a year later, at Alamogordo, New Mexico on July 16, 1945, was the world’s first *known* nuclear explosion. But what experiments were conducted in preparation for that test?[25]

Peter Vogel, an information officer from New Mexico, was sorting through an old box of photo supplies at a church rummage sale near his home in the early 1980s when he found a surprising historic document that apparently came from the nearby Los Alamos Laboratory.

The chart, “*History of 10,000 Ton Gadget*,” showed the eleven steps of a nuclear explosion, from detonation to mushroom cloud. The last entry read, “Ball of fire mushroom out at 18,000 ft in typical Port Chicago fashion.” (“*Gadget*” was the codename for the implosion bomb detonated in the Trinity test.)

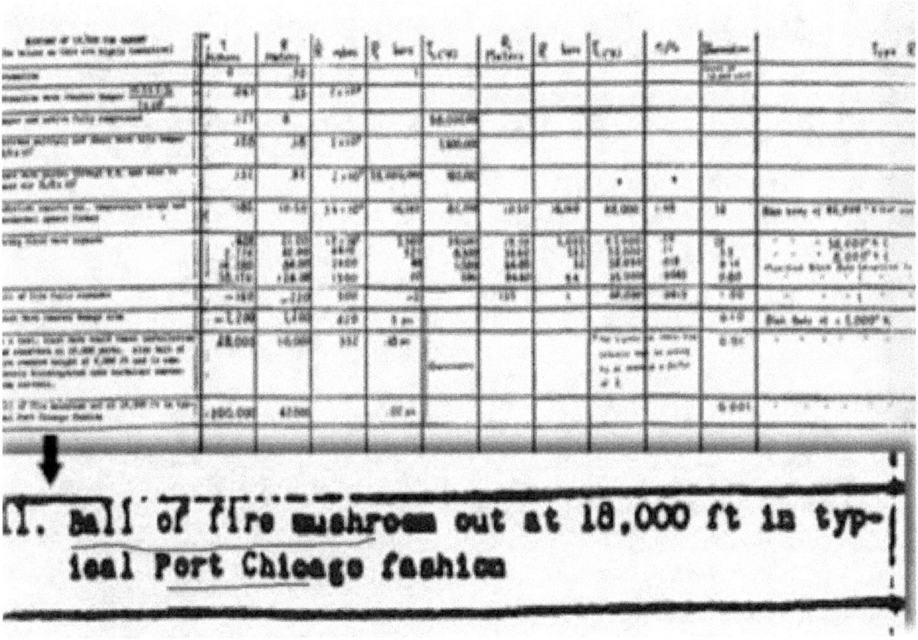
11. Ball of fire mushroom out at 18,000 ft in typical Port Chicago fashion

"History of 10,000 Ton Gadget"

This document sparked Peter Vogel's 35-year independent investigation of the theory that the Port Chicago disaster may have been a nuclear explosion. [26]

"Typical Port Chicago fashion?" Was Port Chicago some kind of model? For what? A nuclear detonation?

Vogel recalled the name of the small town where anti-war protests took place during the 60s, when the nearby Naval magazine shipped arms to Vietnam. But what did Port Chicago have to do with the Manhattan Project?

The question sparked Vogel's curiosity. He began a life-changing research project, presenting his original findings in the Spring 1982 edition of *The Black Scholar.* That issue of the scholarly journal was edited by UC-Berkeley professor (emeritus) Robert Allen, author of *The Port Chicago Mutiny.* His article on the work stoppage and subsequent court martial complemented the cover story, "The Port Chicago Disaster - Was it a Nuclear Explosion?"

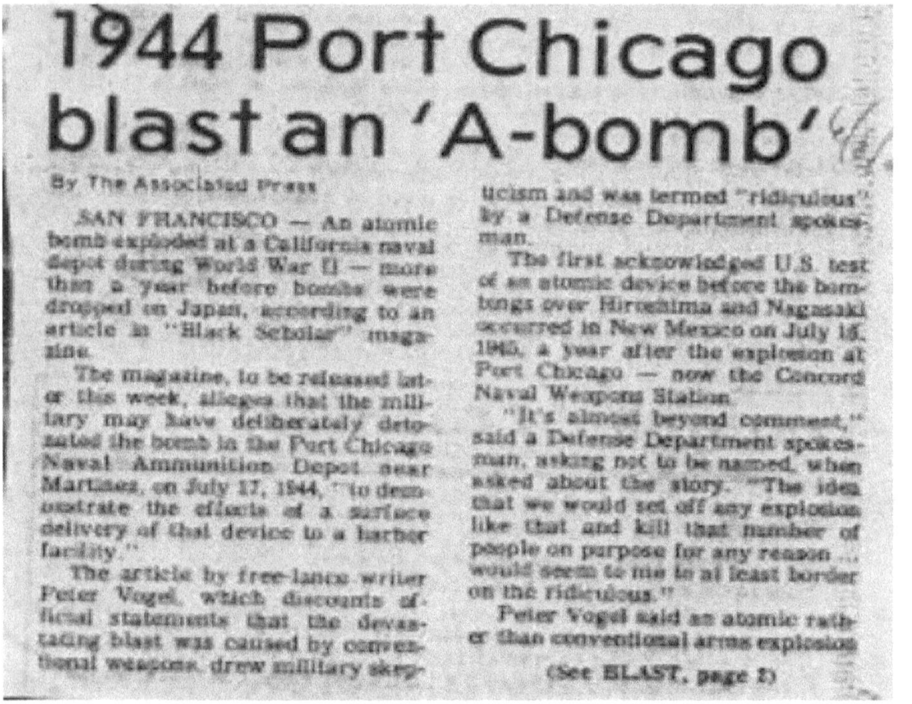

1944 Port Chicago blast an 'A-bomb'

By The Associated Press

SAN FRANCISCO — An atomic bomb exploded at a California naval depot during World War II — more than a year before bombs were dropped on Japan, according to an article in "Black Scholar" magazine.

The magazine, to be released later this week, alleges that the military may have deliberately detonated the bomb in the Port Chicago Naval Ammunition Depot near Martinez, on July 17, 1944, "to demonstrate the effects of a surface delivery of that device to a harbor facility."

The article by free-lance writer Peter Vogel, which discounts official statements that the devastating blast was caused by conventional weapons, drew military skepticism and was termed "ridiculous" by a Defense Department spokesman.

The first acknowledged U.S. test of an atomic device before the bombings over Hiroshima and Nagasaki occurred in New Mexico on July 16, 1945, a year after the explosion at Port Chicago — now the Concord Naval Weapons Station.

"It's almost beyond comment," said a Defense Department spokesman, asking not to be named, when asked about the story. "The idea that we would set off any explosion like that and kill that number of people on purpose for any reason ... would seem to me to at least border on the ridiculous."

Peter Vogel said an atomic rather than conventional arms explosion

(See BLAST, page 2)

"*Port Chicago blast an 'A-bomb'*" - **Associated Press** June 1982

Vogel's theory gained a little traction when it was picked up by the Associated Press in 1982, but it has not been well-received overall. An unidentified Defense Department spokesperson told the Associated Press that the idea was "almost beyond comment" and seemed to "at least border on the ridiculous."

The theory had been mentioned in Wikipedia's feature article on Port Chicago until 2006 when editors – including one who referred to Vogel as "a crackpot" – decided to delete all references to the theory and its author. A lengthy discussion ensued[27], with contributors debating the standard objections. Without any official source to rely on, and with no official investigation, Wikipedia's ultimate decision to delete the theory was based on the standing rule that the crowd-sourced encyclopedia does not publish original research.

"A Story Too Good to Kill"

Historians Lawrence Badash and Richard G. Hewlett wrote a scathing review of Vogel's article, dismissing it as "A Story Too Good To Kill." Other writers have cited their flawed review as proof that the theory has been debunked; but the rebuttal was riddled with fallacies, including speculation, illogical conclusions and straw man arguments.[28]

A Story Too Good to Kill

The "Nuclear" Explosion in San Francisco Bay

LAWRENCE BADASH

University of California

RICHARD G. HEWLETT

History Associates

Sensational stories, like sex scandals, political corruption, and catastrophes have long been the fodder of the news media, whether in the tabloids that one finds at supermarket checkout stands or in respectable daily papers. Not only do such stories enliven our lives, they often feed into a latent suspicion of bureaucratic indifference and ineptitude. When, further, catastrophe is com-

In their 1993 *Sage Knowledge* review of Vogel's theory of a nuclear explosion at Port Chicago, historians Badash and Hewlett referred to his 1982 article in **The Black Scholar** as "*A Story Too Good to Kill.*"

I now turn to detailed analysis, starting with your questions.

1. One thousand seven hundred eighty tons of TNT exploded at one time could, in my opinion, have caused all the effects described in the article (see below).

2. While smoking a cigarette does sound an unlikely reason for the start of the detonation, the actual cause could not have been established after the fact. There are many ways in which carelessness could have produced the disaster. Once started, the detonation wave could have set off both the cargo and the material on the dock as a single explosion. Again it is hard to assess the probability of this happening in this way rather than in a series of explosions, but there is no reason, in my opinion, to find the event improbable, let alone impossible.

3. p. 31 "Ball of fire mushroom out at 18,000 feet in typical Port Chicago fashion". Since this was the largest single explosion anywhere prior to the Trinity test at Almogordo, it was naturally used as a guide to what to expect when the bombs were used. I find the interest of the AEC in the disaster completely explicable for this reason alone.

4. *Capability*. pp. 31-35. This and the next section are for me the most significant parts of the article. I agree that by all accounts the Pu^{239} weapon could not have been exploded much prior to the actual Trinity test, and hence that *if* the Port Chicago disaster were nuclear it would have had to be

Badash and Hewlett cited this letter from physicist H. P. Noyes to RW [**Revolutionary Worker**]. Noyes was responding to an inquiry about Vogel's introductory article on the Port Chicago nuclear explosion theory. His opinion was based on the limited information available to him at the time.)

Al Christman, author of **Target Hiroshima,** was referring to Vogel when he said

> "Parsons' Port Chicago trip was not a major incident in Project Y history despite subsequent sensational attempts by a historical revisionist to make it so. Parsons' appearance at the site, plus the report of a mushroom cloud, contributed to erroneous charges starting in the 1980s that this had been a nuclear explosion and hence the subject of a government coverup. The story resurfaces from time to time despite a clear case to the contrary by historians Lawrence Badash and Richard G. Hewlett, assisted by fifteen other established scholars.[29]

One of those scholars, theoretical physicist H. Pierre Noyes, offered the best point-by-point rebuttal of Vogel's original story; but his arguments, too, are largely based on speculation and opinion.

No investigation

It bears repetition: **there has been no official investigation of the Port Chicago nuclear explosion theory**. By contrast, both the Navy and the National Archives have conducted exhaustive studies of the Philadelphia Experiment, concluding that there is no scientific basis for the claim that a ship was rendered invisible at the Philadelphia Navy Yard in the fall of 1943.

An official investigation is needed. As Christman noted, the story resurfaces from time to time, sensationalized on the internet, for example, under provocative headlines like "Blacks Nuked at Port Chicago."

Badash and Hewlett were concerned that the scandalous story would breed distrust of the government. They were right about that, but, for that very reason, they were wrong in their conclusion that self-respecting scholars should not dignify the story by writing about it. Without any authoritative source to either confirm it or put it to rest, the scandal lingers, veiling the truth. Whatever that is.

A theory too important to ignore

Despite being demonized as a conspiracy theorist and a glory-seeking sensationalist, Vogel continued updating his research for thirty-five years. He consulted archivists and specialists; interviewed Manhattan Project scientists, including Edward Teller, with whom he had once taken a class at UC-Berkeley; searched through innumerable wartime records housed at various agencies, including the Los Alamos National Library. He watched films of the explosion. He studied the damage from the shock wave; the size of the crater in the Suisun Bay; the Wilson condensation cloud; and much more.

Vogel obtained enough documents on the Port Chicago explosion to create his own archive. I contacted him in 2018 with questions about his work and my own research. We shared ideas and exchanged a few documents. That December, the 80-year-old independent scholar shut down the website of his eBook, **The Last Wave from Port Chicago**, still convinced that the detonation at Port Chicago was a secret nuclear test.

Taking Vogel's original theory to another level, this book, **American Nuclear Deception,** presents new, non-technical evidence that the PCNAD was in fact established for that very purpose.

The Last Wave from Port Chicago - Author Peter Vogel

The Last Wave from Port Chicago

by Peter Vogel
Peter.Vogel.US@gmail.com

The result of a 35+ year investigation into the July 17, 1944 explosion at the Port Chicago Naval Magazine.

This web-based book chronicles the total history of the 1944 explosion in California, and connections with Los Alamos and the Manhattan Project.

In December 2018, Vogel shut down the website where he published his eBook, **The Last Wave from Port Chicago**. (An article by that title can be accessed online through a scholarly journal (with a paywall), but it is actually the 1982 article, not the updated eBook, which is now out of print.)

Overcoming the objections

Standard objections to the PCnet are based on opinion, "common sense" and wishful thinking. Each objection is overcome, however, by the evidence.

"Common sense" says:

If there had been a nuclear explosion at Port Chicago, the truth would have come out by now.

The evidence shows that:

Government officials and Manhattan Project leaders used several layers of secrecy to create a permanent cover for the clandestine operation.

"Common sense" says:

The scientists who worked on the Manhattan Project were not evil people. They were determined to stop Hitler from getting the bomb. They would not have used American citizens to test a nuclear weapon.

The evidence shows that:

Scientists studied the means to create the bomb, but military engineers controlled the actual weapon. Very few scientists (if any) would have been involved in planning or carrying out the secret test.

"Common sense" says:

The United States government would not have 'nuked' black sailors.

The evidence shows that:

The PCnet is a serious, evidence-based claim that does not support the sensationalized internet conspiracy theory. The purpose of a nuclear detonation at Port Chicago would have been to test the unprecedented new weapon. There was a war on. Most of the 320 Americans killed at Port

Chicago were expendable military personnel. This includes the 202 Black sailors who died while loading ammunition onto warships as well as the White officers, crewmen and civilians who died in the blast.

BLACK SAILORS NUKED IN ATOMIC BOMB TEST IN PORT CHICAG
The first atomic explosion in the history of the world took place

Without an official investigation of the PCnet, sensationalized rumors persist.

Most of the White personnel who died in the blast were crewmen aboard the SS *Quinault Victory*.

After training for jobs in all ratings at Great Lakes Naval Training Center, some Black sailors loading ammunition at Port Chicago felt they had been "tricked" into signing up.

"Common sense" says:

The United States government would not have exposed its own citizens to the hazards of a nuclear test.

The evidence shows that:

The story of America's Atomic Veterans is one obvious example of U.S. military personnel being used as subjects in hazardous experiments.

Vulnerable (White) civilians have been involuntary subjects of Human Radiation Experiments conducted by the U.S. government in the past.

Race was the main vulnerability factor in the Tuskegee syphilis experiments, sponsored by the Committee on Medical Research under the auspices of the Office of Scientific Research and Development.

U.S. troops participate in *Operation Buster-Jangle-Dog,* the first U.S. nuclear test conducted at the Nevada Proving Ground, 1 November 1951. Sworn to secrecy, Atomic Vets kept quiet about the tests for decades.

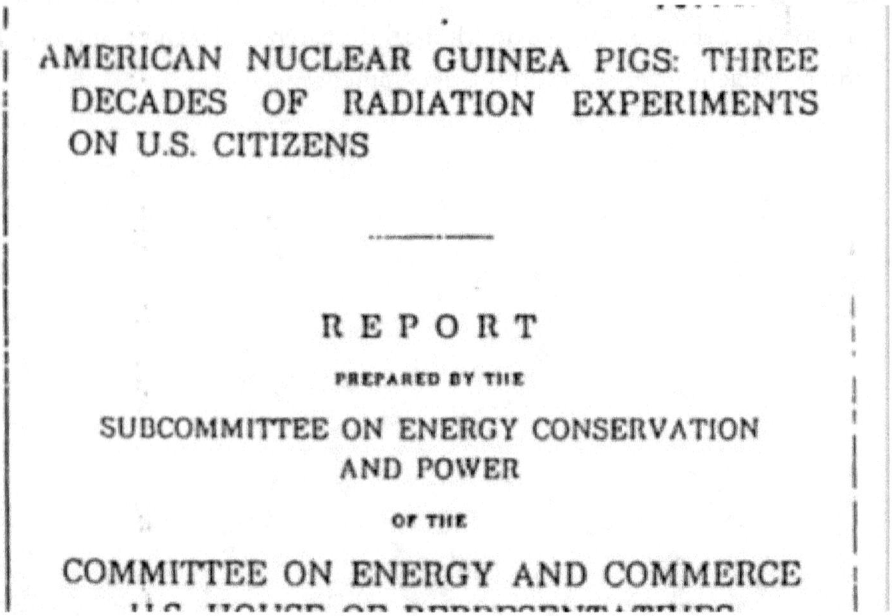

AMERICAN NUCLEAR GUINEA PIGS: THREE DECADES OF RADIATION EXPERIMENTS ON U.S. CITIZENS

REPORT

PREPARED BY THE

SUBCOMMITTEE ON ENERGY CONSERVATION AND POWER

OF THE

COMMITTEE ON ENERGY AND COMMERCE

The U.S. Government sponsored at least 4,000 human radiation experiments between 1944 and 1974.

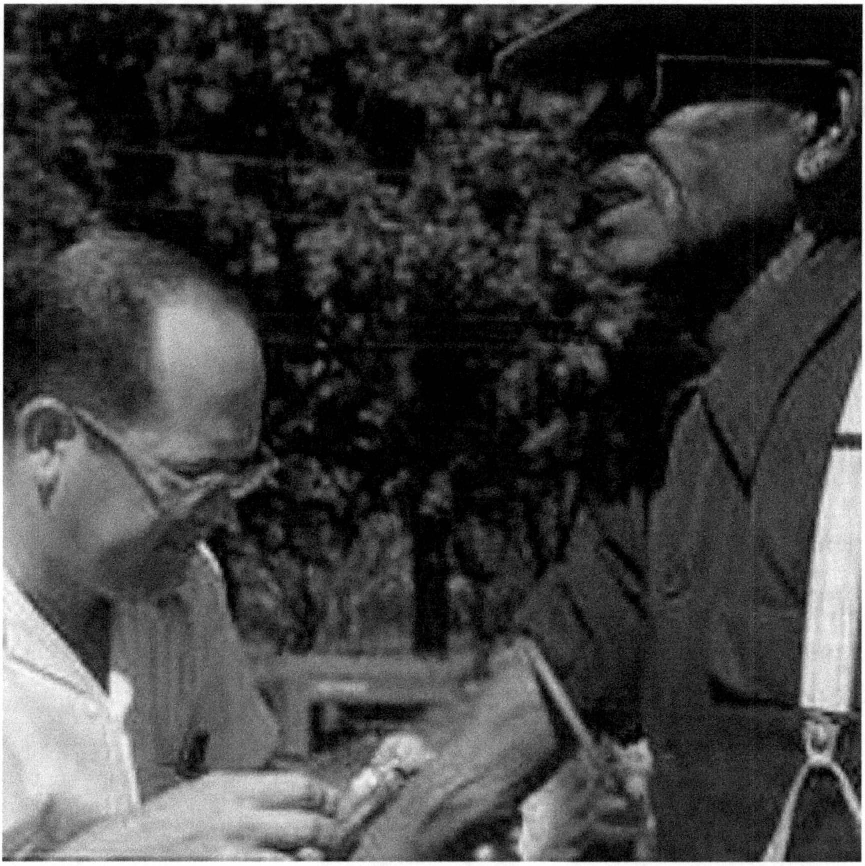

As involuntary subjects of the Tuskegee Syphilis study, Black men were given placebo injections instead of penicillin. The study was sponsored by **the Committee on Medical Research (CMR), an agency administered by Vannevar Bush and the Office of Scientific Research and Development (OSRD).** The buck stopped at Bush's desk. After considering the legalities of the experiment, he signed off.

"Common sense" says:

The government would not have tested a nuclear device at a U.S. Naval facility.

The evidence shows that:

Establishment of the Port Chicago Naval Ammunition Depot coincided with the development of the Manhattan Project, which required a remote, controlled location for conducting hazardous tests of unprecedented magnitude. There was a war on. American men were being killed, American ships and materiel expended in combat zones around the world. The damages at Port Chicago were part of the costs of war, a risky but promising investment in the nation's future.

"Common sense" says:

There would have been radioactive fallout at the site.

The evidence shows that:

There are no visible effects of radioactive fallout; it can be detected only with special instrumentation.

It can also be difficult to determine the source of fallout. For example, any trace of radioactivity found in the area after 1946 could have come from vessels that were contaminated in *Operation Crossroads*, the spectacular tests of atomic weapons against naval ships, and towed back to the Bay area from the Bikini Lagoon.[30]

The spread of radioactive fallout depends on the nature of the explosion. Although the public was never informed, fallout from the Trinity test of July 16, 1945 spread across the U.S.[31]

The government suppressed reports about radioactivity after the bombing of Hiroshima. Even if there was radioactivity at Port Chicago, and even if it had been tracked and evaluated, that information would not have been published.

It would require further research, and possibly a Freedom of Information Act (FOIA) request, to find out if there are any remaining classified documents on the Port Chicago explosion.

Objections overruled

Most (if not all) objections to the PCnet are based on opinion, but the evidence that favors the theory is wide-ranging and compelling. Additionally, research of the PCnet reveals other little-known facts about the bomb and about early nuclear history.

See no evil

Perhaps the most valid objection to the PCnet is one that has *not* been raised. As explained earlier, journalists and historians have not reported on the theory. As a consequence, the general public knows even less about Port Chicago than about nuclear weapons. But if the theory was valid, wouldn't more people know? And if they did, wouldn't they tell? If not, why not?

It is understandable that the story never came out in interviews with people who would have been involved, or in the memoirs of people who would have been eyewitnesses. If anyone knew about an explosion at Port Chicago and did not speak up at the time, they were not likely to divulge that information later. There are a variety of reasons why a person would have kept quiet about such a secret.

Toxic truths: to tell or not to tell

Joseph Rotblat was the only scientist who left the Project after learning that the bomb was not going to be used against Germany. He acted on his conscience, but did not speak out.

Like Daniel Ellsberg, the whistleblower who leaked the Pentagon Papers in 1971, Rotblat later questioned his own motives and devoted the rest of his life to educating the public. Unlike Ellsberg, Rotblat talked his way out of the Manhattan Project with the agreement that he would keep the secret.

The Atomic Vets also kept their secret. Decades passed before they came out with their story. Many suffered major illnesses as a result of radioactive fallout from nuclear tests, but they were told to keep quiet, and they did.

It is unlikely that any of these people were ever aware of the theory that the Port Chicago explosion may have been a nuclear blast. It is reasonable to assume that if the theory was true, someone in a position to learn more would surely have pieced the evidence together. By the same token, the evidence was not published, is not obvious, and runs counter to the traditional history.

Without compelling evidence, people often (incorrectly) assume that because a theory is outlandish, it follows that it is untrue. Nuclear historians aware of the claim have been dismissive of the sensational idea. For one thing, they say, none of the scientists were evil enough to test an atomic bomb at Port Chicago—as if the scientists ever had control of the weapon.

The standard rule applies: great claims require great evidence. But the irony here is that it would take due diligence to obtain great evidence. As in the fable of the *Emperor's New Clothes*, the people close to the king had too much at stake to betray their leader. Even if they did see that he was naked, they sealed their lips.

Lee Basham, a professor of philosophy at South Texas College, explains another reason why people in the know may have kept the secret: some truths are simply too toxic to tell.

Today, the PCnet itself is worthy of study as a fact of history. Whether or not the theory is proven or debunked, it provides a valuable and unique lens for the study of early nuclear history. Like the Geiger counter used to detect radioactive fallout, or like the Luminol used to detect invisible traces of blood at the scene of a crime, the PCnet is a sensitive tool for uncovering unknown evidence about the secret bomb project, facts that have been well-hidden in the traditional history.

Seeing the big picture

The PCnet is supported by many historical facts, including little-known information and some details that may seem minor or unrelated. These data points are best understood in context.

It is important to recognize the bomb not just as a horrific weapon, but as a catalyst for world change. It introduced the nuclear age and, not incidentally, established the "American Century." That, as we will see, may have been the real primary rationale for its creation.[32]

In his 1914 novel, **The World Set Free**, H. G. Wells (the "Shakespeare of science fiction") portrayed the destruction of the world by atomic warfare. The novel ended on a positive note. Alfred Nobel, the creator of dynamite and the Nobel Peace prize, had once hoped that the terrible power of dynamite would help men see the need to put an end to war for all time. Wells had the same hope for atomic weapons.

Through his book, as well as through personal contact, Wells reached several of the people—including Leo Szilard, Winston Churchill, and FDR—who were involved in the decision to create the bomb. These influential men spoke of the atomic bomb not only as a weapon to put an end to war but also as a weapon to influence world affairs in other ways.

Three men—President Franklin D. Roosevelt, Vannevar Bush and William S. Parsons—are of particular interest in the investigation of the PCnet. They would have been the three men most directly involved in the plan to conduct a secret test at Port Chicago. With regard to military science and technology, FDR was influenced by Bush, by Wells, and by the ideas of economist Alexander Sachs.

To assess the PCnet is to examine the motives, the means and the methods of the men who would have carried out the scheme. The urgent need to build the bomb would have driven the need to test it. A full discussion is beyond the scope of this survey of evidence, but this critical element of the PCnet requires careful analysis.

Why would there have been a test at Port Chicago?

Several basic facts are critical background to the development *and use* of the bomb, and to a discussion of why a test was necessary:

- All new technologies are tested. That is what happens in the early research and development phase. These tests show not only whether the technology works, but how. This critical data dictates not only how the technology will be made but how it is to be used. Such information was especially important in the case of the atomic bomb.

- The atomic bomb was a revolutionary technology, a new type of weapon involving rare new hazardous elements. To develop the bomb required new information, new processes, new procedures and new tasks. Unprecedented new problems were involved, including some that arose on the job.

- Traditional history says the world's first nuclear weapon, the *Little Boy* uranium gun-assembly bomb, went into combat without ever having been tested. This dubious premise, one of many in the official narrative, is challenged by the evidence that supports the PCnet—for example, when we consider the "shelving" of its predecessor, the *Thin Man* plutonium gun bomb, on July 17th, 1944—the day of the Port Chicago explosion.

- The weapons used against Japan were in developmental stages; some authoritative people described them as experimental.

- No one knew for certain what to expect from the weapon. In a letter dated September 24, 1944, Parsons protested the idea of a "desert shot": it would not yield the same useful results as a detonation over Times Square, where the human effects would be seen.

- Scientists and engineers preparing for the Trinity test of July 16th, 1945 made bets. Some predicted that the "*Gadget*" would be a dud, the test a fizzle. Others feared the end of the world would come with the apocalyptic blast.[33] They were more confident of the *Little Boy*, which indicates that there must have been some kind of test.

- In their consideration of the options to using the bomb against Japan, Secretary of War Henry Stimson sought the counsel of a select group of scientists, headed by J. Robert Oppenheimer, Director of the Los Alamos Laboratory where the bomb was born. The Scientific Panel could see no alternatives to its use.

- Nuclear tests are scientific experiments; they build on past scientific data. To study new questions, scientists require results from prior tests. The data from the unprecedented blast at Port Chicago explosion was essential and unique.

The benefits of the bomb must be weighed against all of its costs. It follows that the same is true of a test at Port Chicago, which would have been conducted in support of the bomb project.

There have been debates about whether or not the atomic bomb was a legitimate military weapon; but regardless of what side one takes in that debate, it is clear that even before its production, the greatest use of the "exponential weapon" was as a means of influencing international relations.

The war provided a singular opportunity to create and test the bomb. For one thing, secrecy was assured by the war emergency. Social conditions, including widespread racism and the wartime emergency, allowed for extreme and unusual measures. And the hated, unrelenting enemy was a worthy target for demonstrating its power.

The costs

The PCnet must be investigated because it reveals unknown facts about how the creation of the bomb set the foundation of current nuclear policy. In this regard, the PCnet is a cautionary tale.

To Ellsberg, the risk of a nuclear war that destroys the world "has always been an unconscionable risk imposed by the superpowers upon the population of the world." The highest value for those who created the bomb was U.S. military supremacy. Little else mattered, including human lives.

Why the PCnet matters today

Having dispensed with the speculation that the U.S. would not have done it, the PCnet examines why a nuclear weapon would have been tested at Port Chicago. Documentary evidence reveals the details of the scheme and the logistics behind it, showing how the test was conceived, conducted and concealed. The bomb continues to influence international relations. What is less often considered is how the decision to create the bomb, and to keep the project secret from the American public, has had a lasting impact on democracy and on the character of this nation.

Hidden in history

It is one thing to counter the standard objections, but before we begin examining evidence, it should be clear that the PCnet must be taken seriously. We must come to realize that such an operation could indeed have taken place; we must come to understand how it could have been kept secret for so long.

The first answer is circular: several layers of secrecy both made it possible to conduct the clandestine operation and created a permanent cover for it.

The Project, the President and the press

Historians have characterized Franklin D. Roosevelt as always having wheels within wheels. The same thing is true of the Manhattan Project. Although FDR was the man who unilaterally authorized the bomb project, historians have only recently begun to assess the significance of his decision to build the bomb. This omission – from almost all of the many, many books about one of the most important people and one of the most important events of the 20th century – speaks volumes.

FDR was a self-proclaimed juggler who boasted that he never let his right hand know what his left hand was doing. Secrecy was also a defining feature of the Manhattan Project. And there is no automatic declassification for nuclear documents, which have always been "born secret."

It was also possible to conceal the clandestine military operation because

1) It was conducted under the auspices of the top-secret atomic bomb project run by General Leslie R. Groves, a security-minded administrator who controlled the flow of information.

2) It was conducted by a well-connected, powerful and obscure Naval officer, Captain William S. Parsons, who used his rank for cover.

3) It was administered by a secretive powerbroker, Vannevar Bush. As head of the Office of Scientific Research and Development, Bush reported directly to the sphinxlike President. The OSRD was a new, temporary, independent agency in FDR's ever-expanding executive branch.

4) It relied on black budget funding and the cooperation of a few Congressmen who agreed to rubber stamp undisclosed budget requests.

There was no Internet, no WikiLeaks or *Pentagon Papers* during the era of the Manhattan Project. Rather, the national press was governed by a wartime code of "voluntary self-censorship."

Right up until the moment it was dropped on Hiroshima, the bomb was an extremely well-kept secret—a secret kept from the American public. Even the crewmen aboard the *Enola Gay* were astounded by the sight of the mushroom cloud billowing over the city. The only two people who knew what to expect were Capt. Parsons, who served as 'weaponeer', and the pilot, Col. Paul Tibbets. Three weeks earlier, Parsons had flown in the observation plane as a witness to the Trinity test.

The secret was not directed primarily at other nations; Americans were kept in the dark. Citizens whose taxes financed the bomb; those who worked directly for the Manhattan Project and those who worked in other industries that helped build the weapon; the industrialists who ran the factories that supplied the materials; even the Congressmen who authorized the black budget funding – none of these people knew about the bomb until the day it was dropped on Japan.

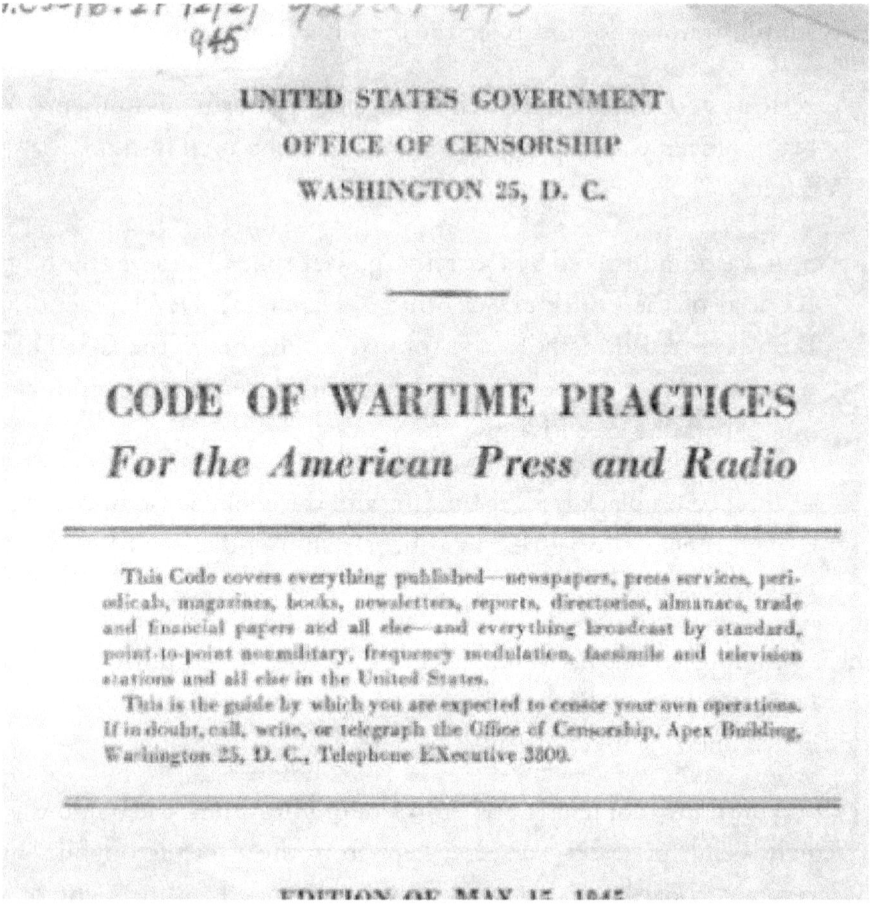

UNITED STATES GOVERNMENT
OFFICE OF CENSORSHIP
WASHINGTON 25, D. C.

CODE OF WARTIME PRACTICES

For the American Press and Radio

This Code covers everything published—newspapers, press services, periodicals, magazines, books, newsletters, reports, directories, almanacs, trade and financial papers and all else—and everything broadcast by standard, point-to-point nonmilitary, frequency modulation, facsimile and television stations and all else in the United States.

This is the guide by which you are expected to censor your own operations. If in doubt, call, write, or telegraph the Office of Censorship, Apex Building, Washington 25, D. C., Telephone EXecutive 3800.

EDITION OF MAY 15, 1945

During World War II, the U.S. press operated under a code of "voluntary self-censorship", under the guidance of Brian Price and the U.S. Office of Censorship.

The Knoxville News-Sentinel

EXTRA

Oak Ridge Has Over 425 Buildings

ATOMIC SUPER-BOMB, MADE AT OAK RIDGE, STRIKES JAPAN

Story of Secret Officially Told

Two Views of Giant Oak Ridge Production Plant

Until Truman announced the atomic bomb attack on Japan, August 6, 1945, most Manhattan Project employees lived in one of three "secret cities", had no outside contact and no idea what they were working on.

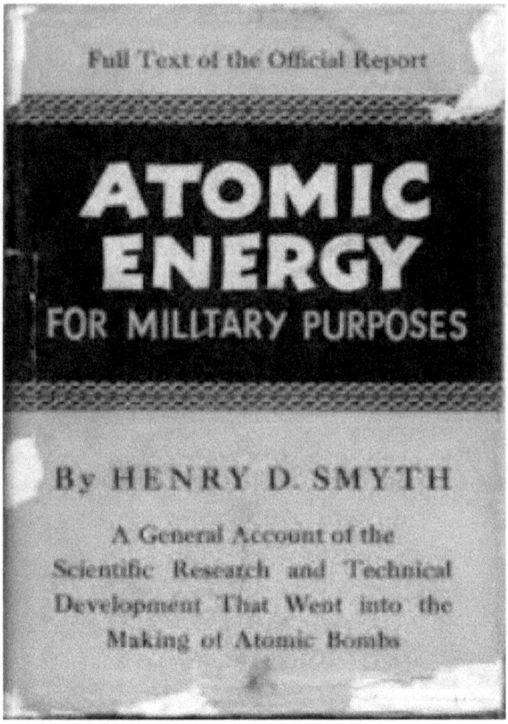

The Government released **The Smyth Report**, the first official history of the atomic bomb, on August 12, 1945, a week after the bomb was dropped on Japan. The official title of the report was **Atomic Energy for Military Purposes.**

Groves commissioned the first official history of the bomb. He wanted **The Smyth Report** written as if from his viewpoint.

[redacted] OFFICIAL USE ONLY

SUMMARY

1. Introduction. - In general (1-1) this guide has been written to present the purpose of the Manhattan District History, the broad considerations involved, form of arrangement and presentation desired. The purpose of the history (1-2) is to serve as a source of historical information for War Department officials and other authorized individuals. Accordingly, the viewpoint (1-3) of the writer (a) should be that of General Groves and the reader (b) should be considered as a layman without any specialized knowledge of the subject who may be critical of the Department or the project. The subdivisions (1-4) of the history

Groves established guidelines for **The Smyth Report**: "*. . . the reader should be considered as a layman without any specialized knowledge of the subject who may be critical of the Department or the project.*"

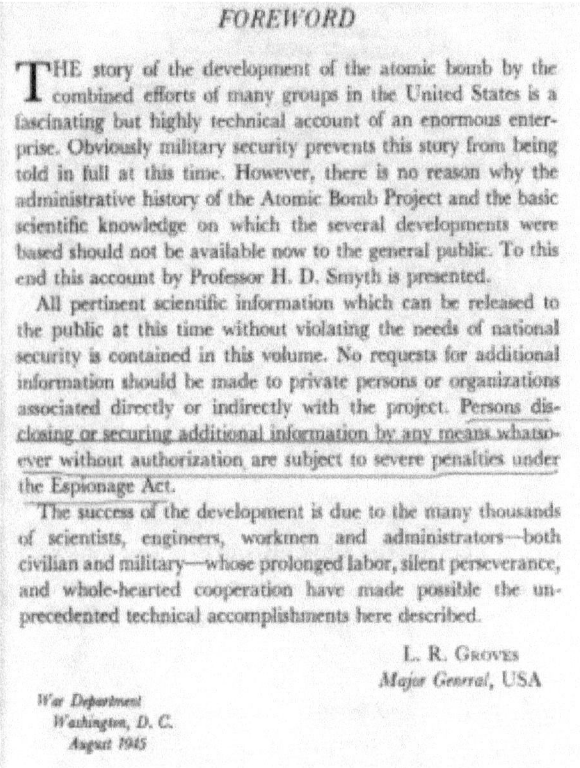

FOREWORD

THE story of the development of the atomic bomb by the combined efforts of many groups in the United States is a fascinating but highly technical account of an enormous enterprise. Obviously military security prevents this story from being told in full at this time. However, there is no reason why the administrative history of the Atomic Bomb Project and the basic scientific knowledge on which the several developments were based should not be available now to the general public. To this end this account by Professor H. D. Smyth is presented.

All pertinent scientific information which can be released to the public at this time without violating the needs of national security is contained in this volume. No requests for additional information should be made to private persons or organizations associated directly or indirectly with the project. Persons disclosing or securing additional information by any means whatsoever without authorization are subject to severe penalties under the Espionage Act.

The success of the development is due to the many thousands of scientists, engineers, workmen and administrators—both civilian and military—whose prolonged labor, silent perseverance, and whole-hearted cooperation have made possible the unprecedented technical accomplishments here described.

L. R. Groves
Major General, USA

War Department
Washington, D. C.
August 1945

"*Persons disclosing or securing additional information by any means whatsoever without authorization are subject to severe penalties under the Espionage Act*." L. R. Groves, Major General, USA – Preface to **The Smyth Report** [underlining added]

The Smyth Report acknowledged interest in the military use of the bomb. This original historical account neither opened with nor emphasized the Einstein letter:

> The announcement of the hypothesis of fission and its experimental confirmation took place in January 1939, as has already been recounted in Chapter I. There was *immediate interest in the possible military use* of the large amounts of energy released in fission. At that time *American-born nuclear physicists were so unaccustomed to the idea of using their science for military purposes*

> that they hardly realized what needed to be done. Consequently the *early efforts both at restricting publication and at getting government support* were stimulated largely by a small group of foreign-born physicists centering on L. Szilard and including E. Wigner, E. Teller, V. F. Weisskopf, and E. Fermi. [*emphasis added*]

The Smyth Report also noted some of the important and varied roles of Alexander Sachs.

> Within the next few weeks a number of people concerned, particularly *Sachs*, urged the importance of greater support and of better organization. Their hand was strengthened by the Columbia results (as reported, for example, in a letter from *Sachs* to General Watson on May 15, 1940) showing that the carbon absorption was appreciably lower than had been previously thought and that the probability of carbon being satisfactory as a moderator was therefore considerable. *Sachs* was also active in looking into the question of ore supply. On June 1,1940, *Sachs*, Briggs, and Urey met with Admiral Bowen to discuss approaching officials of the Union Miniere of the Belgian Congo. Such an approach was made shortly afterwards by *Sachs*. [*emphasis added*]

According to **The Smyth Report**, the Einstein letter was *not* a warning to FDR about the possibility of a German bomb. Sachs used the letter signed by Einstein as part of a fuller dossier in support of his presentation on October 11th and 12th, 1939, when he "explained to President Roosevelt the desirability of encouraging work in this field."

Decades later, though, this information is usually excluded from historical accounts.

The Smyth Report sold out quickly because it was the first source of public information about the bomb, but it was highly technical. Groves made certain that the report emphasized the scientific aspects of the bomb project. He deleted references to ordnance and engineering.

The need to dress up the story for public consumption came later, when increasing criticism of the bomb led Project leaders to worry about public perception. They were concerned that in the future, misguided educators might teach American students bad things about the bomb. And what if the President decided to release more public information?

Groves decided to hire William Laurence, a former reporter for the *New York Times*, as a publicist. Other Manhattan Project leaders joined the campaign to sell the "Atomic Crusade." It is no accident that today, almost every account of the bomb's creation begins with the misleading legend of the Einstein letter.

23. I also believe that a paragraph should be inserted which pays tribute to the voluntary cooperation of the 2,000 daily newspapers, the 11,000 weekly newspapers and the ______ radio stations in the country. Many of them knew our secret and kept it a secret. Many others could have guessed it from the Censorship directive, but did not guess.

24. In the radio speech, I would emphasise,near the top, the race with the Germans. I would elaborate on this in the Secretary's statement. Both would be for the purpose of calling attention to our plight if the Germans had succeeded in developing the atomic bomb. Selling for the future, especially the Congress, also is involved.

WILLIAM A. CONSODINE,
Lt. Colonel, C.E.

"...In the radio speech, **I would emphasize, near the top, the race with the Germans.** I would elaborate on this in the Secretary's statement. Both would be for the purpose of calling attention to our plight if the Germans had succeeded in developing the atomic bomb. **Selling for the future, especially the Congress, also is involved.** [emphasis added]

- William A. Consodine, Lt. Colonel, C.E.[34]

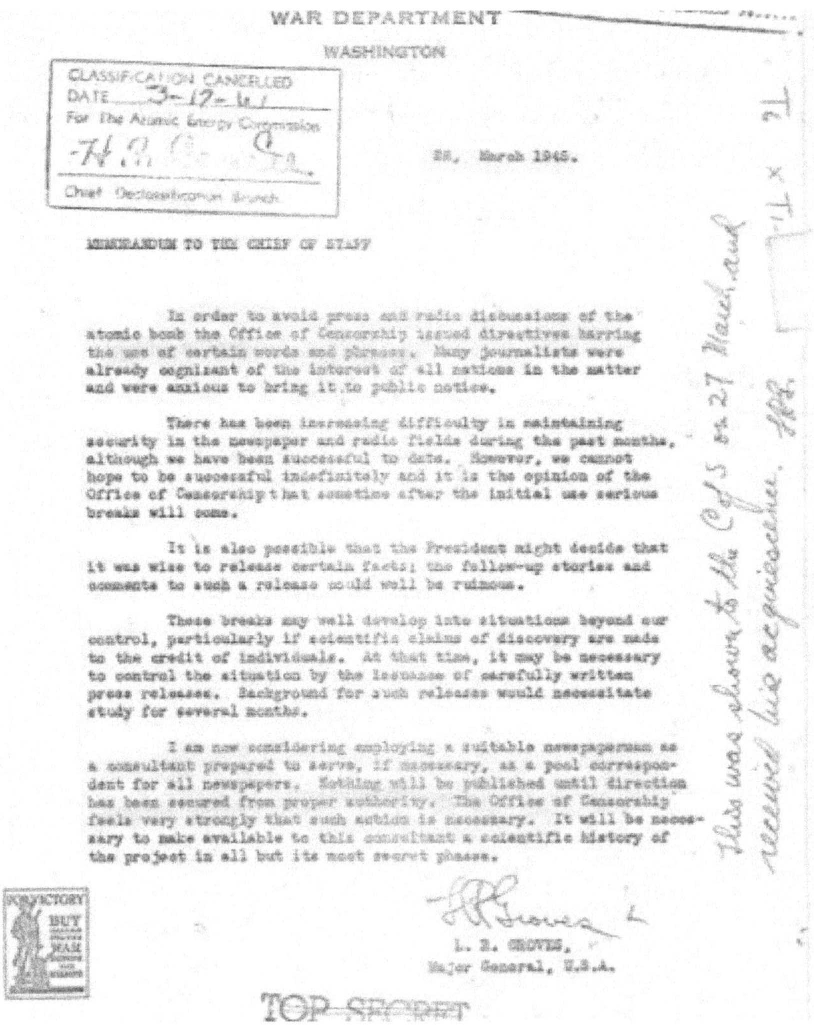

WAR DEPARTMENT
WASHINGTON

CLASSIFICATION CANCELLED
DATE 3-17-61
For The Atomic Energy Commission

Chief Declassification Branch

26, March 1945.

MEMORANDUM TO THE CHIEF OF STAFF

In order to avoid press and radio discussions of the atomic bomb the Office of Censorship issued directives barring the use of certain words and phrases. Many journalists were already cognizant of the interest of all nations in the matter and were anxious to bring it to public notice.

There has been increasing difficulty in maintaining security in the newspaper and radio fields during the past months, although we have been successful to date. However, we cannot hope to be successful indefinitely and it is the opinion of the Office of Censorship that sometime after the initial use serious breaks will come.

It is also possible that the President might decide that it was wise to release certain facts; the follow-up stories and comments to such a release would well be ruinous.

These breaks may well develop into situations beyond our control, particularly if scientific claims of discovery are made to the credit of individuals. At that time, it may be necessary to control the situation by the issuance of carefully written press releases. Background for such releases would necessitate study for several months.

I am now considering employing a suitable newspaperman as a consultant prepared to serve, if necessary, as a pool correspondent for all newspapers. Nothing will be published until direction has been secured from proper authority. The Office of Censorship feels very strongly that such action is necessary. It will be necessary to make available to this consultant a scientific history of the project in all but its most secret phases.

L R Groves

L. R. GROVES,
Major General, U.S.A.

This was shown to the C of S on 27 March and received his acquiescence. LRG

TOP SECRET

Early public information releases set the assumptions and misleading parameters for the faulty history of the bomb.:[Highlights added.] Note **middle paragraph:**

'It is also possible that the President might decide that it was wise to release certain facts; the follow-up stories and comments to such a release could be ruinous."

- L.R. Groves to Chief of Staff, March 26, 1945

The *story* of the Einstein letter was part of a calculated campaign to sell the bomb to the American public. If nuclear secrecy was for the purpose of national security, then as far as the Manhattan Project was concerned, the American public was enemy number one. Even after the use of the bomb, when the Project itself was no longer secret, someone still had something (nontechnical) to hide.

The "Einstein letter" story

Albert Einstein did not write the famous "Einstein" letter. This foundational fact bears repeating. The false textbook narrative is deliberately deceptive; yet almost every article about the history of the bomb opens with the *misleading myth* that Einstein wrote a letter to the President warning him about the possibility of a German bomb.

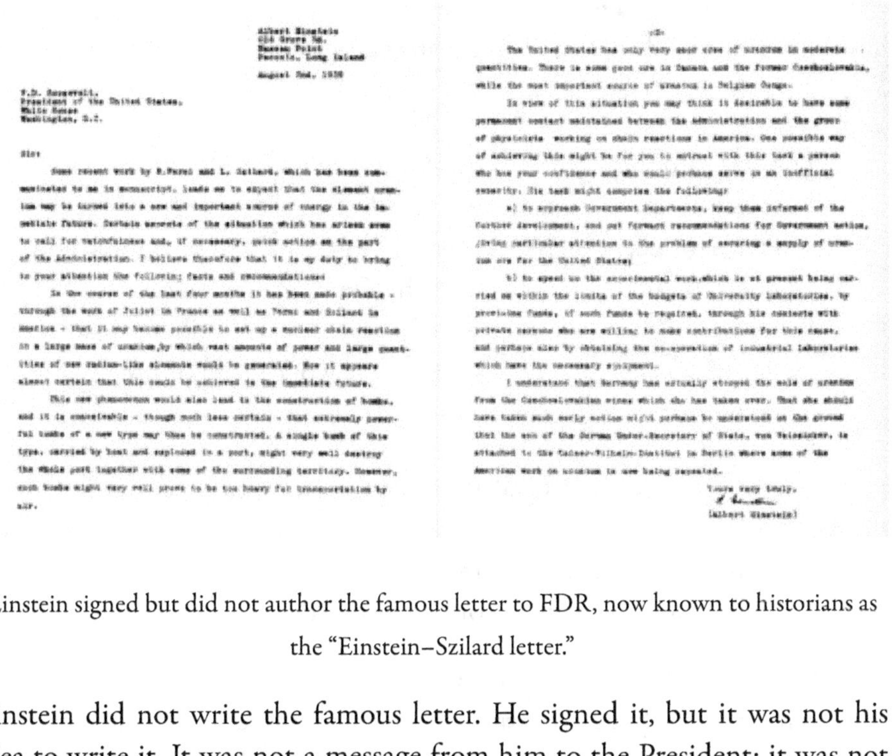

Einstein signed but did not author the famous letter to FDR, now known to historians as the "Einstein–Szilard letter."

Einstein did not write the famous letter. He signed it, but it was not his idea to write it. It was not a message from him to the President; it was not intended to express a concern he had about a German bomb. So, who *did* write it? Whose idea was it? What does the letter really say? What was its real purpose?

And, most importantly, who created the false impression that Einstein wrote the letter? Why did they perpetuate the narrative that places Einstein at the forefront of the bomb's history?

The Einstein letter was a sales ploy. After the war, the national press and the public bought the noble and misleading story that President Roosevelt ordered the bomb project because (they believed) Albert Einstein, the beloved humanitarian genius, wrote a letter warning the President that Germany was working on the bomb.

Almost all historical accounts follow that misleading premise. They also maintain that the letter was a "warning". In fact, the single reference to Germany in the closing paragraph makes no explicit reference to the potential threat of a German bomb. A more explicit interpretation is that the government should act fast to secure uranium from Europe in case Germany cut off all access.

> I understand that *Germany has actually stopped the* <u>*sale*</u> *of uranium from the Czechoslovakian mines which she has taken over*. That she should have taken such early action might perhaps be understood on the grounds that the son of the German Under-Secretary of State, von Weizeacker, is attached to the Kaiser-Wilhelm-Institut in Berlin where some of the *American* work on uranium is now being *repeated*. [*emphasis added*]

The letter did not prompt FDR to order any effort to determine the status of Germany's uranium research, which would have been a reasonable immediate response to such a warning, especially if the decision to invest in the risky, hazardous and costly project was based on fear of a German bomb.

The first attempt to notify the government was not a warning to the President or to the State Department; it was an unsuccessful outreach to the Navy. The next appeal for government support for the ongoing research by the emigrant scientists now working in the U.S. was the letter to FDR. (It is quite likely that the Navy consulted the President before responding to the scientists' inquiry.)

Several writers contributed to the letter. They had at least two different motives. For one thing, they foresaw an opportunity to exploit nuclear energy for domestic purposes after the coming war, but the more immediate concern was the possibility that "... extremely powerful bombs of a new type may thus be constructed." There was a war on, and the side that got the bomb first would be at an advantage. That kind of talk would appeal to a shrewd president, encouraging him to support uranium research.

There are additional little-known sides to the Einstein letter story. Both Sachs and Bush give different versions of the events leading up to the bomb. In his testimony before the November 1946 Senate hearings, Sachs claimed that he informed FDR about the discovery of nuclear fission and sold him on the idea of pursuing the bomb. That conversation took place in the spring of 1939, months before the Einstein letter was drafted.[35]

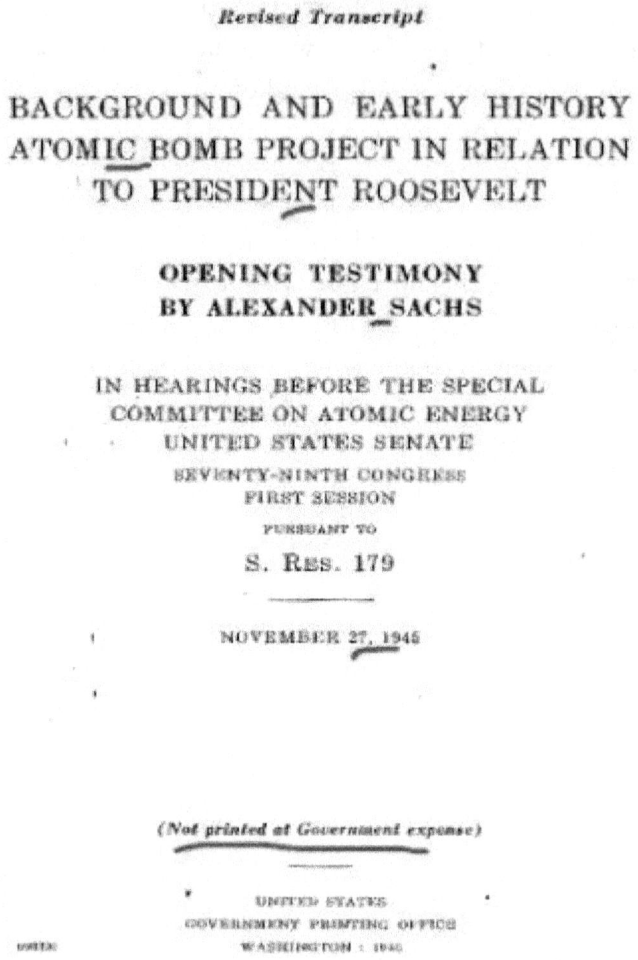
Revised Transcript

BACKGROUND AND EARLY HISTORY
ATOMIC BOMB PROJECT IN RELATION
TO PRESIDENT ROOSEVELT

OPENING TESTIMONY
BY ALEXANDER SACHS

IN HEARINGS BEFORE THE SPECIAL
COMMITTEE ON ATOMIC ENERGY
UNITED STATES SENATE
SEVENTY-NINTH CONGRESS
FIRST SESSION
PURSUANT TO
S. RES. 179

NOVEMBER 27, 1945

(Not printed at Government expense)

UNITED STATES
GOVERNMENT PRINTING OFFICE
WASHINGTON : 1946

Wall Street economist Alexander Sachs was the first witness called to testify before the November 1945 *Senate Special Committee Hearings on Atomic Energy.*

Several sources indicate that Sachs and Einstein had a prior acquaintance. In fact, Sachs reported that he had had an opportunity to speak with Einstein and other leading scientists about the status of atomic research while on tour in Europe, well before the January 1939 announcement of nuclear fission.[36]

Sachs helped several European scientists immigrate to the U.S. Further research would be necessary to show whether Einstein was one of them. Sachs did report meeting with Frank Aydelotte, Einstein's boss at the Institute for Advanced Studies (IAS) at Princeton and, according to this report, Einstein had already agreed to sign a letter to FDR.

Ironically, although he claimed to have been the person who recommended that the scientists obtain Einstein's signature, Sachs later came to resent the superficial history that emphasizes the role of the scientists.

One of the challenges of historical research is interpreting the evidence, weighing one document against others, one source against others, one account against others. History is revealed in stories. As in any other circumstance, interpretation depends upon the interpreter's previous knowledge and point of view.

Historians have an additional responsibility to avoid (or at least admit) any personal biases that may influence how they tell the story. And, of course, when history is revised, each new account of an event either supports or detracts from the prevailing narrative.

Eminent historian Barton J. Bernstein warns scholars that Sachs is an unreliable source.[37] Yet his papers are vital documents, in some cases are the only primary source available. This is especially true, of course, with regard to his conversations with FDR. For whatever reasons, historians cite some of his anecdotes while overlooking others.

It may be that Sachs *appears* unreliable simply because his contemporary documents tell a different story than the contrived narrative based on the myth of the Einstein letter. It is also possible that he was a glory-seeker who wanted to get his name up in letters for all time. In either case, it is also clear that the men who created the traditional story had high stakes in how it was told.

For that reason, it is important to study and evaluate Sachs, treating his claims the same as those of any other witness. To accurately evaluate new historical evidence, the traditional account must be treated as one possible *version* of the story; then the actual historical events must be reviewed in light of all the evidence.

It might also be worthwhile to study the events surrounding Einstein's immigration, as well as the simultaneous creation of the IAS. Physics was a new field in the 1940s, and the relatively small community of physicists was centered in Europe (in fact, in Germany). Just how did it happen that the best minds in physics and other key fields came together from around the world in time to develop the bomb project in America at the start of World War II?

The fact that many of the emigrant scientists were Jewish is one part of the story. According to Sachs, he and President Roosevelt understood that a German bomb was not likely. The fact that Hitler was opposed to their "Jewish science" further weighs against fear of a German atomic bomb as the catalyst for the American project.

The Bush influence has been downplayed in traditional history, but it is of particular interest to the PCnet.

In May 1938, Vannevar Bush left his post as Vice President at the Massachusetts Institute of Technology (MIT) and moved to Washington, where he became president of the Carnegie Institution of Washington (CIW). He leveraged this position to get close to the White House. In 1940, when FDR made him tsar of science, Bush was able to push for greater use of science in the military.

Although skeptical at first about the feasibility of an atomic bomb, Bush was in on the plan to develop it almost from the start. His prominent role is one of the facts concealed behind the standard Einstein story.

Perspective makes a great difference. A 1959 report on *Basic Research in the Navy* is one of the rare sources to note that "this stage of history" was influenced by Bush as well as by the Einstein letter.

> "...At almost the same time Dr. Vannevar Bush, Chairman of the National Advisory Committee for Aeronautics, began to

> formulate plans for a National Defense Research Committee to "... coordinate, supervise and conduct scientific research on the problems underlying the development, production, and use of mechanisms and devices of warfare..."[38]

As noted by the *PBS History Detectives*, the role of Bush and Conant in managing the Manhattan Project did not become known until 1958, when

> Richard G Hewlett, the official historian of the US Atomic Energy Commission (AEC) opened an old filing cabinet under a stairwell and pulled out a remarkable collection of wartime correspondence files of two men: Vannevar Bush and James B. Conant ... to this day, the role of Bush and Conant is one of the lesser-known stories of the nuclear arms race.[39]

The Einstein story overshadows the secret role Bush played in the secret history of the secret project. Bush himself refuted the significance of the Einstein letter and minimized the scientists' impact on the decision to create the weapon.

The emigrant scientists sought Government support for their uranium research, which needed to be conducted on a larger scale after the discovery of nuclear fission. They worked with Alexander Sachs, the Wall Street Banker and Presidential advisor, who had his own reasons for influencing FDR to pursue the bomb. Vannevar Bush, who sought to promote science for military and government use, resisted the bomb at first but came to recognize it as a technology whose time had come.

When "revisionist" historians began taking a second look at Truman's decision to use the bomb, they did so without revisiting the bomb's original creation story. In recent years, historians have finally begun to reevaluate the importance of FDR's decision to create the bomb. His motives and actions are the ultimate key to the Port Chicago nuclear explosion theory.

As we will see, the famous Einstein letter supports the idea of a nuclear explosion at Port Chicago in several ways. As for the bomb project, though, the letter sold the ennobling idea that the bomb was the brainchild of the beloved humanitarian scientist, a pacifist and a genius. In fact, the author of the letter was Leo Szilard, an emigrant from Hungary, Einstein's friend and colleague. He conceived the idea and enlisted Einstein's support. Again, that version of the story comes closer to the account given by Sachs, who insisted that it was his idea to address the letter to the President.

The PCnet provides another viewpoint, another premise, for assessing the letter. Bush speaks to a challenge that is otherwise overlooked: Did emigrant scientists really have that much influence over such a momentous decision? Even if their letter had been the spark for FDR's interest, what actually caused the dynamic flame that led to the $2 billion project (1940s dollars)? What did it have to do with Alexander Sachs, and what did Sachs have to do with it?

A providential opportunity

Contrary to popular history, when Sachs met with FDR on the evening of October 11th, 1939, he was no mere courier hand-delivering Einstein's letter. There is no historic indication that FDR ever read the letter, which was one of three documents in a more complete dossier in support of Sachs' presentation, which resumed the next morning. (There was also a letter from Sachs himself and a scientific report prepared by Szilard.)

Sachs was known for his wordiness and flowery language. This time, he was also convincing. *He did* warn the President: FDR must not make the same mistake Napoleon Bonaparte had made. The great emperor may have determined the course of history and sealed his own doom when he turned down Robert Fulton's offer to build a fleet of submarines.

Sachs kept a lot of records. The extent to which they are true and accurate requires analysis, but they are available; other primary sources are rare. Historians caution researchers not to rely too heavily on Sachs, but he was both the man in the room and the man with the pen.

Every source is important; they all shed some light on events and they all have deficits. Bush, for example, reported that FDR only spoke about the bomb with a few people; he did not include Sachs on the list because he did not know about him. After all, FDR prided himself on never letting his right hand know what his left hand was doing.

Stories about the Einstein letter often close with a convenient quote from Sachs, where FDR turned to his aide, Admiral Edwin "Pa" Watson, and said, "This requires action." Historians do *not* report that Sachs later complained about the simplified press coverage that overemphasized the famous letter.

They also omit a most telling comment that casts the bomb's creation story in a whole new light. According to Sachs,

> "The President summed up his conclusion and conviction by saying, 'I almost think there is something providential about the

> challenge and the opportunity put up to the American democracy and people.[40]

President Franklin D. Roosevelt was a self-proclaimed 'juggler'. He made it clear: to win the war, he was willing to lie, if necessary.

Lie? About what? What else was the bold and visionary risk-taker willing to do, and why?

Roosevelt and the writer, who were reflecting on that extraordinary piece of unhistorified history in the summer of 1940, so overladen with darkness for the democratic cause, was the following: Although Nazi Germany had the lead in atomic research, the monolithic character of the regime and its

- 10 -

systemic exclusion of the play of free and pluralistic thought would militate against the progress of atomic research just because it was so new and just because the required condensation of the normal historical course for the evolution from an idea into a finished instrument would call for pluralistic novelties and resourcefulness. The President summed up his conclusion and conviction by saying, "I almost think there is something providential about the challenge and the opportunity put up to the American democracy and people."

"I almost think there is something providential about the challenge and the opportunity put up to the American democracy and people." - FDR to Sachs, Spring 1939 – Sachs Papers, Atomic.07, FDR Library

The truth about Truman

On July 11th, 1944, FDR finally announced that he was going to run for a fourth term. A few days later, he revealed his real choice for a running mate. But prior to that time, he had dissembled, telling Party leaders "I hardly know Truman."[41] It is unlikely, though, that FDR really did not know the man who headed the powerful Truman Committee, the Senator who appeared on the March 8, 1943 cover of **Time Magazine** as one of the most important people in Washington.

Most sources say President Harry Truman did not know about the bomb until after his inauguration.

But what if he did? That would shed a very different light on the strange circumstances surrounding his nomination at the Democratic National Convention when Party bosses shut down the voting to prevent the re-election of Vice President Henry Wallace. The next day, after effective overnight political maneuvering, Truman won an upset victory.

The election took place four days after the Port Chicago explosion. FDR missed the convention; he was traveling to the West Coast. It is possible, of course, that all of these somewhat related historic events just happened to occur around the same time.

By the same token, it would have taken an elaborate scheme to pull off a nuclear test at Port Chicago, and an additional set of clever plans to keep it concealed for decades. These stories seldom overlap in historical accounts, which would have made it easier for the clandestine operation to be covered up with some sophisticated sleight of hand and several layers of secrets.

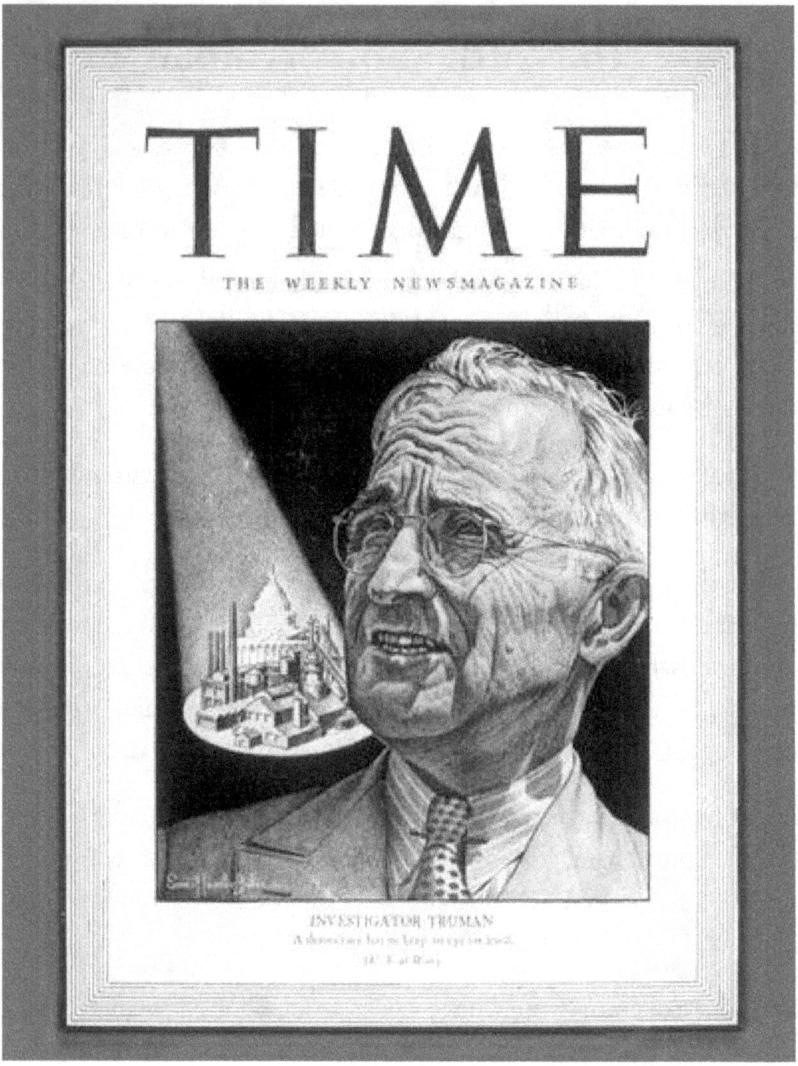

Investigator Truman - "A democracy has to keep an eye on itself." - (U.S. at -War)

"For a Congressional committee to be considered the first line of defense—especially in a nation which does not tend to admire its representatives, in Congress assembled— is encouraging to believers in democracy. So is the sudden emergence of Harry Truman, whose presence in the Senate is a queer accident of democracy, as the committee's energetic generalissimo."

Billion-Dollar Watchdog, **Time Magazine,** *March 8, 1943*

If there was a cover-up, the story might begin to unravel when we pull at any one of several strings.

For instance, the PCnet provides a different way to review the under-examined story of how President Harry Truman, the "accidental President", came to preside over one of the most momentum decisions in history.

While most historians insist that Truman knew little or nothing about the bomb until after his inauguration, it is entirely understandable that few (if any sources) show conclusively that President Roosevelt never told Truman about the Manhattan Project. First of all, how does anyone know what FDR did or did not tell Truman? It is difficult under any circumstances to determine what someone did *not* say to someone else in a private conversation.

Besides, there were other people, including Vice President Henry Wallace, who could have conveyed the message on behalf of FDR (thus affording Truman the "plausible deniability" that accompanies the notion that the President withheld the information; especially considering that both Marshall and Stimson warned him to steer the Truman Committee clear of the Hanover construction site.)

Other than his own secretiveness, historians offer no logical reason why FDR would have withheld the information from Truman once they became running mates. The President knew he was ill, and he knew that if he was unable to complete his term, responsibility for the bomb would fall on his predecessor. There is no way to know precisely what he was thinking, but it was more within his character to have devised an intriguing scenario (and perhaps lead people to think that he had not spoken to Truman about the bomb) than to put his legacy at risk by handing it over to someone who might dismantle it or handle it poorly.

What if Truman did know?

There are other indications that Truman had at least enough information about the Manhattan Project to have raised the issue with the President.

(Or to have kept shrewdly quiet about it.) He was the popular head of the powerful committee that investigated defense spending. Both Secretary of War Henry Stimson and General George Marshall had cautioned him to steer the Truman committee clear of the mammoth and mysterious construction project taking place at Hanover, Washington (where the Manhattan Project produced plutonium.)

Truman may have been the person referenced by Arthur Compton, one of the Manhattan Project leaders, who explained the success of the secrecy in his post-war article, "The Atomic Crusade."

> "...But the real explanation for the tightness of the secrecy was that when by accident someone learned what was going on, he was awed by its significance. A simple suggestion was enough to make him understand that "here is something I must not tell. The safety of the world may depend on the tightness of my lips." He had himself become a part of the crusade."[42]

As seen, left to right: President **Franklin D. Roosevelt**, Vice-President-elect **Harry S. Truman,** and Vice-President **Henry Wallace**.

Finally, undergirding the likelihood that Compton was referring to Truman, there is the clear statement in **The Smyth Report** by Henry D. Smyth, the first official history of the Manhattan Project, which was commissioned and controlled by the Project's Director, General Leslie R. Groves. For some reason, Smyth states that Senator Truman "*. . . had been aware of the project and its magnitude.*" (Perhaps this statement is not to be interpreted literally; or it may be that **The Smyth Report** itself is inaccurate or misleading.)

5.34. Since the earliest days of the project, President Roosevelt had followed it with interest and, until his death, he continued to study and approve the broad programs of the Military Policy Committee. President Truman, who as a United States Senator had been aware of the project and its magnitude, was given the complete up-to-date picture by the Secretary of War and General Groves at a White House conference immediately after his inauguration. Thereafter the President gave the program his complete support, keeping in constant touch with the progress.

The **Smyth Report** says "...President **Truman**, who as a United States Senator **had been aware of the project** and its magnitude..." [emphasis added]

It is hard to know quite what to think about Truman and his ascendancy to the highest office in the land. Contradictions beg to be resolved:

- According to most sources, Truman was an unlikely candidate for Senator, let alone for President.

- Yet contemporary issues of *Time Magazine* identified the popular Senator leading the Truman Committee as one the most influential people in Washington.

- A 1992 article in *American Heritage Magazine* reported that when Democratic Party bosses looking to replace Vice President Henry Wallace brought up the popular Senator from Missouri, FDR protested "I Hardly Know Truman."[43]

- By contrast, in a 2012 article titled, "How Truman Led the Charge in Congressional Oversight," Robert M. Poole reported that

"Truman's skillful handling of the panel, which managed to be critical of the Roosevelt administration without being malicious, influenced FDR's decision to tap him as a running mate in 1944—a fateful choice that landed the Missourian in the Oval Office a few months later."[44]

● In a June 24, 2020 interview sponsored by the FDR Library and the WWII Museum, Clifton Truman Daniel, a Truman scholar, insisted that his grandfather knew nothing about the bomb project before becoming President.[45]

● According to a Senate history of the Truman Committee, "The committee's work prompted criticism from officials in the Roosevelt administration. When Congress declared war on Japan on December 8, 1941, critics at the White House and within Congress called for the committee's dissolution.

● In a November 2015 retrospective on "*Gen. Marshall and the Atomic Bombing of Japanese Cities*" published in *Arms Control Today*, Barton J. Bernstein noted that the Truman Committee actually *sheltered* the FDR administration, and later Truman, himself from close scrutiny.

"Initiated early in the war under then-Senator Truman (D-Mo.), the panel had helped insulate the government's policymaking on defense issues from public scrutiny. When Truman, elected vice president in 1944, became president in April 1945, that Senate committee, maintaining its ways from the Roosevelt period, continued to help protect the chief executive and his bipartisan national security system from intrusive scrutiny on the conduct of the war.[46]

Historians resist using counterfactuals (stories that present events in ways we know they did not happen) to study history. On the other hand, the question "what if" is useful for exploring questions that remain unanswered.

If — as these sources seem to indicate — Truman did know about the bomb, why do so many people say otherwise? *If* he did know, what was the origin of the myth that he did not? What could be the point of not simply saying so? And if he did not know, why didn't he? After all, the popular and highly effective Truman Committee had oversight over defense spending during the war.

In his speech accepting the nomination to Vice-President (a job he allegedly had resisted), Truman made it clear that he would be maintaining FDR's program. Since the general consensus is that Truman did not know about the bomb at the time, historians have not had the weapon in mind when assessing his speech.

From the PCnet perspective, however, Truman's insights regarding ongoing discussions concerning a forthcoming decision that might bring about a sudden end to the war are notable. He seemed to sense that he might soon have to rely on the judgment of people who were already well-conversed with the topic. He might have been thinking of the likes of James Byrnes, FDR's advisor, who later became Truman's Secretary of State, and was very influential when it came to the bomb.

> "The end of hostilities may come suddenly. Decisions that will determine our future for years, and even generations to come, will have to be made quickly. If they are made quickly and wisely by those who have had years of experience and the fullest opportunities to become well informed with respect to our national and international problems, we can have confidence that the next generations will not have to spill its blood to rectify our mistakes and failures.[47]

Note in particular his reference to the lasting implications for how wars would be conducted in the future. FDR made similar veiled comments in his final, undelivered speech. Such comparisons are beyond the scope of the traditional and revisionist narratives of early nuclear history. Such unorthodox questions arise when viewing history through the lens of the PCnet.

The ongoing "atomic crusade"

Discrepancies in the official narrative, as well as omissions, unanswered questions and misleading statements, have led historians to revisit the early history of the bomb in recent years.

Wartime secrecy was maintained after the war by a coordinated public information campaign to sell the bomb to the American people. This early history was deliberately designed to cast the horrific weapon in a favorable light and to forestall Congressional hearings.

The March 8, 1943 *Time Magazine* article, *Billion-Dollar Watchdog*, quotes Truman saying,

> "The goal of every man on the committee is to promote the war effort to the limit of efficiency and exertion. It doesn't do any good to go around digging up dead horses after the war is over, like the last time. The thing to do is dig this stuff up now, and correct it. If we run this war program efficiently, there won't be any opportunity for someone to stir up a lot of investigations after the war—and cause a wave of revulsion that will start the country on the downhill road to unpreparedness and put us in another war in 20 years. . .."

Then, as the war ended, President Truman announced the bomb as "the greatest scientific achievement in history."

Wondrous — and horrific — as it was, American got the new weapon first (preventing a dictator like Hitler from using it to conquer the world). The weapon, we are told, had to be used against Japan (in order to win the war and save lives). The weapon that ended the war and saved lives was God's gift to the U.S.

Almost as FDR had predicted, the providential weapon provided an opportunity for America, and a challenge to democracy.

The bomb brought about a revolutionary change in the world order.

Many Americans, including publishers and even military leaders, denounced the use of the bomb, calling it barbaric. Some detractors changed their tune a bit in 1949, however, when Russia tested its first nuclear bomb. The Cold War was on, providing further rationale for the ongoing development of the U.S. nuclear arsenal.

In a sense, the tail was wagging the dog. Despite postwar talk of international control and limitation of nuclear weapons, U.S. policymakers had already determined it was necessary to stockpile nuclear weapons.

In 1946, as the United Nations met to consider international control, the U.S. was preparing for *Operation Crossroads*, which would demonstrate the power of the U.S. bomb for all the world to see.

Operation Crossroads

On paper, the call for tests of atomic bombs against naval ships came from Senator Brien McMahon (D-Conn) and Admiral Ernest King. In fact, though, *Operation Crossroads* was the fulfillment of an idea Parsons first detailed in an independent report he wrote on July 4th, 1944. Two weeks before the Port Chicago explosion, Parsons was already making plans for an even grander experiment, one that would be witnessed by the world.

Three tests were planned. Dr. Stafford Warren, Chief Medical Officer of the Manhattan Project, warned against conducting *Test Baker*, the second test, an underwater detonation. To observers the first test, *Test Able*, was anti-climatic. It did not live up to the hype of pre-test publicity. The third, *Test Charlie*, was canceled after *Test Baker* proved to be far more hazardous than anticipated.

Warren had accurately predicted excessive hazards from radioactivity. Parsons responded in typical Parsons' fashion: the test was more important than the potential harm it could bring to others.

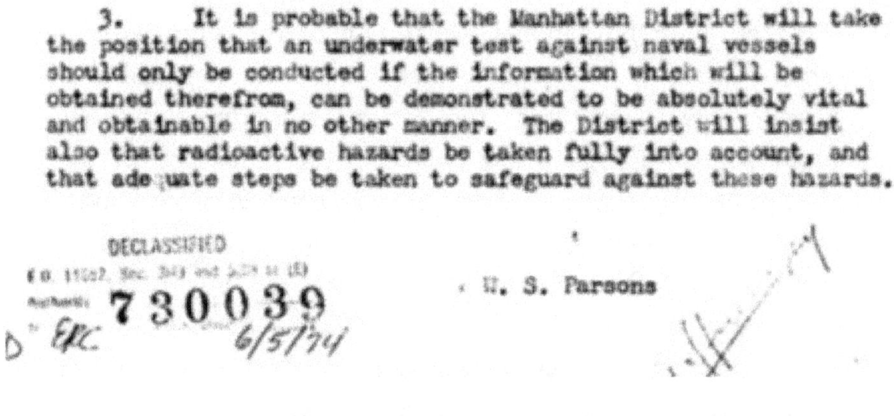

3. It is probable that the Manhattan District will take the position that an underwater test against naval vessels should only be conducted if the information which will be obtained therefrom, can be demonstrated to be absolutely vital and obtainable in no other manner. The District will insist also that radioactive hazards be taken fully into account, and that adequate steps be taken to safeguard against these hazards.

DECLASSIFIED

730039

6/5/74

W. S. Parsons

Despite warnings that great hazards were likely, Parsons insisted on conducting the *Operation Crossroads* tests:

"*It is probable that the Manhattan District will take the position that an underwater test against naval vessels should only be conducted if the information which will be obtained therefrom, can be demonstrated to be absolutely vital and obtainable in no other manner . . .*"

Oppenheimer was one of many who objected to the tests, saying that they would be of limited use, of great hazard, and that there were safer and more effective ways to obtain the needed information. He felt so strongly about it that he risked Truman's wrath by asking to be excused from the President's *Crossroads* advisory panel. (Truman did not let him off the hook.)

In June 1947, the Joint Chiefs of Staff Evaluation Board submitted a report on *Operation Crossroads* titled "*The Evaluation of the Atomic Bomb as a Military Weapon.*"

Some of their recommendations are evident in current nuclear policy. For example, the Board recommended that the U.S. must continue **stockpiling** weapons:

> (3) In the absence of absolute guarantees of abiding peace, the United States has no alternative but to continue the manufacture and stockpiling of weapons of nuclear fission and to carry on continuous research and development for their improvement and improvement in the means of their delivery.[48]

The Board also recommended that the U.S. revise its stand on a **first strike**:

> (5) There must be national recognition of the probability of surprise attack and a consequential revision of our traditional attitudes toward what constitute acts of aggression so that our armed forces may plan and operate in accordance with the realities of atomic warfare. Our policy of national defense must provide for the employment of every practical means to prevent surprise attack. **Offensive measures will be the only generally effective means of defense, and the United States must be prepared to employ them before a potential enemy can inflict significant damage upon us.** [emphasis added]
>
> For this policy to be effective, the **President** should be empowered to make the instant, **unilateral decision** to make an

offensive nuclear attack, in case another nation was *contemplating* an attack on the U.S.:

> (6) National defense requirements of the future are only those of the past; any aggressor must be overcome with superior force. But, where in the past, the duty of the President, as the Commander in Chief has been restricted (before formal declaration of war) to action only after the loss of American lives and treasure, it must be made his duty in the future to defend the country against imminent or incipient atomic weapon attack.

> (7) What constitutes an "aggressive act" or "imminent" or "incipient" attack it is the responsibility of the Congress to define and to redefine, from time to time, so that it may draft suitable standing orders to the commander in chief for prompt and effective atomic bomb retaliation should another nation ready an atomic weapon attack against us.

Maintaining the stockpile would require **ongoing nuclear tests**:

> (17) Dominance in the ability to wage atomic warfare, the loss of which might be fatal to our national life, can be retained only by unflagging effort to hold that leadership in science and engineering which made the atomic bomb possible. A vital part of the national defense must be not only a program of scientific and engineering research and development in every field involved in bomb production and tactical use, but in basic science as well. Such a program requires periodic tests of atomic weapons.

FDR's bomb, as it turned out, disproved one of his most famous sayings. Now there was more to fear than fear itself. Regretfully, peace could only be enforced by "**universal fear of the atomic bomb**..."

> (18) A peace enforced through fear is a poor substitute for a peace maintained through international cooperation based upon agreement and understanding. But until such a peace is brought

> about, this nation can hope only that an effective deterrent to global war will be a universal fear of the atomic bomb as the ultimate horror in war.

The Board recommended, among other things, that it should be made the **President's duty**, upon consultation with the Cabinet, **to order atomic bomb *retaliation*** when such "retaliation" was necessary to *prevent* or *frustrate* an atomic weapon attack upon us.

Again, national security required **ongoing, full-scale tests of nuclear weapons, whenever and wherever necessary**.

> (10) That tests of atomic weapons be held whenever necessary in the research and development of such weapons or the tactics of their use or the training of personnel.

And further

> 7. No development of any weapon or skill in its use can be achieved without periodic tests. The phenomenon of nuclear fission precludes the explosion of an atomic weapon of less than critical mass, and hence **all tests must be "full scale"**. [emphasis added]

The nature of the atomic bomb made it most effective against **urban targets**. The bomb was, to a great extent, a psychological weapon. Fear of the bomb, then, would be most effective in industrialized nations with large cities, where people were aware of the threat—e.g., the United States.

> 12. Even a cursory examination of the characteristics of the American people and of the cultural and material fabric of their national life invites the conclusion that this nation is much more vulnerable to the psychological effects of the bomb than certain other nations.

The Board promoted a "revolutionary change in military mental attitudes." The stakes were high:

> The possible penalty of a failure to retain dominance in the development of the atomic bomb and of the strategy and tactics of its use is so great that it must serve as a constant incentive to the best thought and effort of our military planners.

These excerpts from early nuclear policy contain three clues to the PCnet:

> 1) The idea that **all tests must be "full scale"** suggests that the PCNAD would have been a suitable proving ground because it would provide a location where a full-scale test could be conducted in a secret, combat-like setting.
>
> 2) The ongoing need for ever-escalating tests after the war underscores the need for a test during the early research and development of the bomb, when the unknowns were greatest.
>
> 3) Parsons – the man in charge of the development of the weapon – was an experimental officer who promoted early and frequent testing during the R&D stages of the Manhattan Project.

During the war, the men administering the Manhattan Project saw no alternative to building and using the bomb (that is, demonstrating its effectiveness in a combat situation.) To build the unprecedented new technology inherently involved testing it.

After the war, there were no alternatives to the constant pressure to build more weapons and conduct more tests.

Neither were there any alternative voices of authority to counter the influence of people like Parsons. The first "atomic admiral"— Technical Director of *Operation Crossroads*—was the first person named in the Board's acknowledgements:

> "In the preparation of this report the Board availed itself of the services of Rear Admiral W. S. Parsons, USN, Chairman, JOINT CROSSROADS COMMITTEE and found his advice and assistance of great value."

The one consistent voice of influence, from the origin of the bomb through the establishment of early nuclear policy, is that of William S. Parsons, the man whose July 1944 vision was fulfilled in 1946 by what he described as "the largest laboratory experiment in history, conducted by a Task Force of 40,000 men, 4,000 miles west of the West Coast."[49]

One way to trace the early history of the bomb, and the development of U.S. nuclear policy, is to follow the career of William S. Parsons. His obscurity goes a long way toward explaining how the history of the bomb became so murky. It also explains how a test at Port Chicago could have been conducted and hidden for decades.

The first atomic admiral

At the time he died on December 5, 1953, William Sterling Parsons was a highly decorated officer, the first "Atomic Admiral."

Rear Admiral William Sterling Parsons, USN, [reproduction of color painting], oil on canvas board; by Charles Bittinger; 1946. /www.history.navy.mil

As biographer Albert Christman points out in **Target Hiroshima**, the list of Parsons' personal and professional contacts reads like a *Who's Who* among top scientists, military leaders and politicians of the 1940s. To Christman,

> Deak Parsons was a new kind of warrior in the midst of a weapons revolution. He fought his battles in the laboratory and on the proving grounds. When he went into combat he did so with new and radical weapons he had helped to create.[50]

Parsons was the only person to either witness or otherwise participate in seven of the first eight nuclear detonations. He was a brilliant and innovative military man, well-respected by his colleagues and commanding officers, admired by those he taught and commanded.

Christman acknowledges that Parsons' story provides key insights for understanding the creation of the bomb and the birth of the nuclear age. As he explains, the decades that passed between the time of the events and the publication of Parsons biography provided a useful delay that permits greater perspective.

> In particular, the radiation hazards and the moral issues of the atomic bomb are seen in a more discerning light now than when still under the shadow of World War II. **These issues, *plus revisionist efforts to reshape nuclear history*, make it all the more important to examine the circumstances that brought the atomic bomb into being and into combat use.**[51] [emphasis added]

Parsons is best known as the Navy Captain who armed the *Little Boy* atomic bomb aboard the *Enola Gay* as the special B-29 headed from Tinian Island to Japan; but that heroic feat hardly represents the numerous unique and important ways in which he served his country. He had a hand in each of the three new technologies—radar, the proximity fuze, and the atomic bomb—that transformed the U.S. military into the supreme power it became during World War II.

Ironically, Parsons' contributions have been overlooked because of his key role in the Manhattan Project. The omission began with the post-war campaign to explain the atomic bomb to U.S. citizens. For whatever reasons, the officials who shaped that history found it necessary to limit information about the Navy, engineering, and the Ordnance Division.

Parsons' only biography, published in 2014, more than makes up for the oversight. Christman's biases are obvious and useful. By presenting Parsons as a superlative military leader *who could do no wrong*, Christman provides many close details otherwise unavailable. By filtering the facts from the author's opinions – a good practice whenever critical thinking skills are required, and especially necessary when assessing a controversial idea like the PCnet – an impartial observer with enough additional information can discern important information about the events that occurred and the decisions made.

As a key contributor to the progress of military science, as leader of one of America's most successful military undertakings during the war, and then as an influential military leader after the war, Parsons was an exemplary officer. Still, his actions and statements take on different significance when viewed through the lens of the Port Chicago explosion.

The qualities that made him a military genius also make Parsons the number one person of interest in the Port Chicago nuclear explosion theory (PCnet). Consider, for example, Parsons' youthful experiments laying metal on railroad tracks to see what would happen when a train ran over them. To Christman, this was a sign of initiative and curiosity. From that same activity, the PCnet recognizes a fearless young innovator who was content to put himself and others at risk in order to satisfy his curiosity and accomplish his exceptional goal.

The blurb for **Target Hiroshima** says, "For better or worse, Navy captain William S. "Deak" Parsons made the atomic bomb happen." The creation of the atomic bomb was a rare opportunity, made possible in part by a rare officer whose personality and power helped shaped the nuclear age. Parsons' story – his power, position, and personality, as reflected in his statements, actions and attitudes – provide some of the most compelling evidence in support of the PCnet.

Racism, pragmatism and the PCnet

Captain Parsons was a results-oriented experimental engineer. His feeling upon seeing the mushroom cloud over Hiroshima was one of awe, along with a profound sense of responsibility. He had witnessed the Trinity test and knew what the "Japs" were in for; but, he said, he felt no particular emotion about it.

To some historians, this attitude may indicate a trace of "subtle racism". Whether or not this is perceived as a racist attitude depends on the person making that judgment. In fact, racism is not one set of ideas or attitudes but rather a complex concept, manifested in many different ways, which vary with individual persons and under various circumstances.

Racism can be pragmatic. As such, it can facilitate or be a means toward achieving other goals. Paul Williams addresses this fact in **Race, Ethnicity and Nuclear War.** He cites historian Andrew J. Rotter, who noted several reasons for the decision to use the weapon. In addition to the impact it would have on the war itself, and on future relations with the Soviet Union, there was the simple need to use the expensive weapon in a real-world setting, both to test the effects and to demonstrate its power. Williams found these conclusions more credible than the assertion that American racism led the USA to use atomic bombs against the Japanese – although undoubtedly their perceived sub-humanity meant that US policymakers had fewer reservations than if a Germany city had been identified as the first concrete target."[52]

Animosity toward the Japanese was widespread and acceptable among Americans during World War II. (And vice versa: as John Dower points out in **War Without Mercy**, government propaganda demonizing "the enemy" was rampant in Japan as well as in the U.S.)

This "otherization", as Dower terms it, makes it easier to treat human beings as non-human. In times of social pressure, as in war, otherization narrows moral lines. Seeing the enemy as less than human makes it harder for otherwise decent people to take the higher road that values all of human life.

Put another way, racism makes it easier to make decisions and take actions that would ordinarily be out of bounds. That was, in fact, one of the arguments people used to justify using the bomb against civilian populations: the use of the atomic bomb, they argued, was no worse than the firebombing of cities. The "moral threshold" had already been crossed when World War II became a "total war," with unconditional surrender as a goal.

This does not suggest, however, that there is no moral threshold. Rather, such rationale provides an excuse for ignoring the subtle guidance of human conscience.

"Oh, what a tangled web we weave . . ." The poem by Sir Walter Scott about the complications of a lie could be applied to the top-secret Manhattan Project. Whatever else is said about the Manhattan Project and nuclear secrecy, it is time to acknowledge that secrecy and deception are no strange bedfellows. The one almost invariably accompanies the other. In the case of nuclear history, the Government's secrecy has resulted in **American self-deception**. The resulting hubris is reflected in the vitriol exchanged by the conflicting groups who fought over the Smithsonian Institution's controversial plans to commemorate the 50th anniversary of the Hiroshima bombing.[53]

There is no end to the heated and unresolvable debates about whether the bombing was justified, necessary, right. Whether or not it ended the war is a somewhat two-sided argument. Whether or not it saved lives and, if so, how many, is less easily proven. What is clear, though, is that the bomb continues to be a divisive point among Americans. And this is the case despite the fact that the weapon not only killed at least 100,000 people in one blow, it targeted civilians. What does it say about American sensibilities that, eighty years later, we still argue about offering an apology?

Perhaps we will eventually take a hint from the American Psychological Association, which issued an apology for its biased treatment of African

Americans. The American Nurses Association has done likewise. Other similar organizations have recognized and apologized for the error of their ways, and many are at least talking about what they can do to redress the harm done. At an elaborate and moving ceremony conducted by indigenous Canadian tribes on July 25, 2022, Pope Francis apologized for the role of the Catholic Church in the historic oppression of Indian students at residential schools.

In each of these three cases, and other historic settings, including South Africa's reckoning with apartheid, **truth** was the key to forgiveness and reconciliation.

Perhaps the most important, multi-level lesson to be learned from the use of the bomb, and from the PCnet, is how easy it is to cross an invisible moral line. How natural and necessary it then becomes to cover up abhorrent action with distortions and deception. How readily we deceive ourselves and accept creative narratives used to excuse unacceptable behavior. And how crucial the truth is when we finally decide to pursue reconciliation.

Responding to a letter from a clergyman concerned about American values and the ethics of the bomb, President Truman expressed his own related concerns, but explained that "When you have to deal with a beast you have to treat him as a beast."*[54]*

Speaking on America's behalf, Truman exemplified the cognitive dissonance that comes from rationalizing our own loathed behavior. Interpreted through the PCnet, which contemplates the value of all human life, is "regrettable" when we feel we have to reduce ourselves to the state where we treat people whom *we deem to be beasts* as beasts, the people we deem to be "others" as others. (And then we are stymied by the increasing rate of American polarization and violence.)

Truman's letter, and the rationale *still offered today* by many who defend the use of the atomic bomb, equates Japan's surprise bombing of Pearl Harbor and their unconscionable maltreatment of American prisoners of war with the two sudden, surprise nuclear attacks that destroyed two cities, immediately killing an estimated 170,000 civilians and causing the deaths and multi-generational suffering for an inestimable number of others.

This inequitable tit-for-tat does *not* take into account the incredible and varied ongoing costs to America, including the ongoing threat of nuclear holocaust and the psychological impact that Robert J. Lifton and Greg Mitchell refer to as "psychic numbing in their insightful book, **Hiroshima in America**."[55]

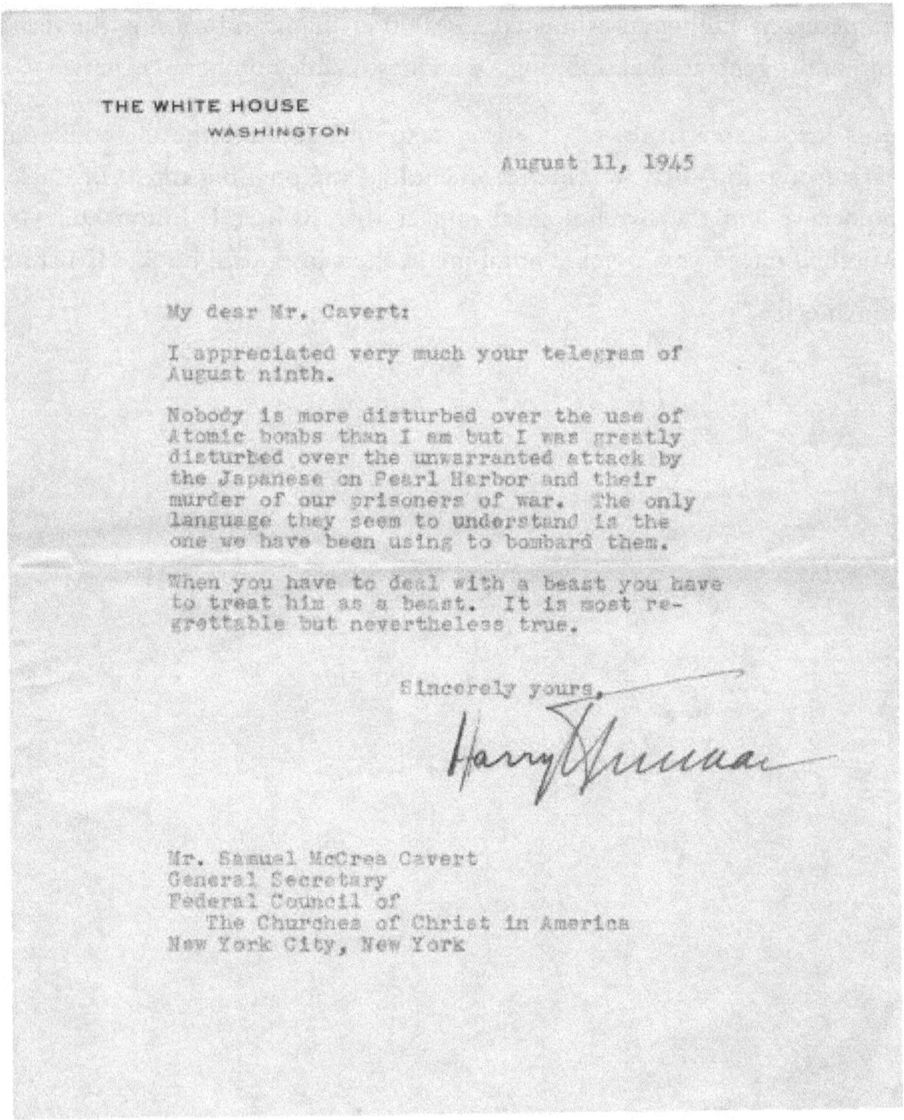
THE WHITE HOUSE
WASHINGTON

August 11, 1945

My dear Mr. Cavert:

I appreciated very much your telegram of August ninth.

Nobody is more disturbed over the use of Atomic bombs than I am but I was greatly disturbed over the unwarranted attack by the Japanese on Pearl Harbor and their murder of our prisoners of war. The only language they seem to understand is the one we have been using to bombard them.

When you have to deal with a beast you have to treat him as a beast. It is most regrettable but nevertheless true.

Sincerely yours,

Harry Truman

Mr. Samuel McCrea Cavert
General Secretary
Federal Council of
The Churches of Christ in America
New York City, New York

"Nobody is more disturbed over the use of Atomic bombs than I am . . ."

- President Truman to Mr. Samuel McCrea Cavert, Federal Council of the Churches of Christ in America, August 11, 1945[56]

President Truman, who presided over the decision to use the bomb, had a plaque on his desk with the phrase "the buck stops here." Truman accepted responsibility for the decision to use the bomb against Japan, although others (including General Groves) pointed out the built-in momentum, concluding that Truman's decision was more or less made by default acceptance of the status quo.

At any rate, historians still debate 'the decision' to use the bomb. Alex Wellerstein, author of **Restricted Data,** points out that

> Rather than one big "decision," the atomic bombings were the product of a multitude of many smaller decisions and assumptions that stretched back into late 1942, when the Manhattan Project really got started.[57]

One decision led to another. Among the chain of events leading to the bomb, various people made various decisions for various reasons: the decision to conduct uranium research . . . the decision to seek government support for the research . . . the decision to fund uranium research (with a black budget) . . . the decision to study the feasibility of a bomb . . . the decision to produce the bomb . . . to produce a different type of bomb . . .

The chain of decisions is important to the PCnet. If there was a nuclear test at Port Chicago, someone made an early decision to test the weapon and to do all the things necessary to make that possible. The moral threshold would have been crossed at some point (or points) during the decision-making process.

Just as Parsons' wartime investigation of a nuclear torpedo led to *Operation Crossroads*, the test of a nuclear weapon at Port Chicago would have paved the way for its use against civilians in Japan. The more nonhuman the victims, the more expendable they would have been.

Ultimately, though, the bell that sounded the alarm of atomic destruction at Hiroshima and Nagasaki has tolled for everyone. Eventually, the (mostly White) sailors who manned the decks at *Operation Crossroads* and America's (mostly White) Atomic Veterans became direct victims of those very early decisions to create and test the bomb. Today's fears about nuclear war result from the same series of decisions.

In **The Age of Hiroshima**, Wellerstein cites Malloy:

> "...many of the options for how to use the first atomic bombs were determined by technological choices made earlier by weapons designers who were far removed from the strategic discussions. Cities were targeted, in part, because the bombs that were built were not very good at doing much else."[58]

One of the important lessons of the PCnet, and of the bomb itself, is that racism can facilitate actions that might otherwise be considered immoral. Racial prejudice or hatred may or not even be part of the motivation; rather, the perceived need to take an unsavory action—to test a nuclear weapon, for example—can itself be the impetus for actions that are racist by virtue of their impact on a group of people of a particular race, or members of some other "other" group. Again, race is only one form of 'otherization', alongside nationality, gender, religion, economic status, ethnicity and other bases for war.

Simply put, vulnerable people make good victims; and "others" are vulnerable people. When necessary, people can be "otherized" for the sake of creating a needed target group.

Senator Burton Wheeler may have been on the wrong side of the noble history of the "Good War" when he complained that FDR would "plow every fourth American boy under."[59] Fear that young soldiers will be used as "cannon fodder" is an inherent part of war. On the other hand, as World War II approached, African American leaders sought the opportunity to serve the country and thereby secure equal rights for the race.

Three Black leaders – A Philip Randolph, head of the Brotherhood of Sleeping Car Porters, T. Arnold Hill of the Urban League and Walter White, Secretary of the NAACP – met with FDR and Navy Secretary Frank Knox on September 27, 1940.[60] To their dismay, FDR opposed integration of the Navy, a position that became clear three weeks later in an announcement by his press secretary, Stephen Early, suggesting that the Black leaders had approved the standing segregation policy. After an uproar, FDR apologized for the misunderstanding "on both sides".

By the summer of 1942, though, the President had changed his mind. He began insisting that the Navy must be open to Negroes around the same time that he okayed plans for the Manhattan Project. This under-explored historic coincidence is of interest to the PCnet.

"BuNav . . . invent something"

FDR, a visionary internationalist, foresaw the significant role the U.S. would play in the coming war. He prepared for U.S. involvement in a variety of ways, including strengthening the military. He had already begun building up the Navy as Assistant Secretary of the Navy during World War II, and that work intensified as the war approached.

But integration of the Navy was not among FDR's plans at the outset of World War II.

Whether by pure coincidence or as a coordinated set of incidents, FDR's change of heart regarding the integration of the Navy in January 1942 coincided with the establishment of the Manhattan Project. Because of the coincidental timing – which becomes apparent only with the study of the PCnet – the first Negro sailors were available to staff the Port Chicago Naval Ammunition Depot when it opened in December 1942.

As noted earlier, on September 27, 1940, FDR and Secretary of the Navy Frank Knox met with three Black leaders, A. Philip Randolph of the Brotherhood of Sleeping Car Porters; T. Arnold Hill of the Urban League, and Walter White, Secretary of the NAACP. In addition to jobs for Negroes in defense factories, they were pushing for integration of the Navy.[61]

Knox was adamant; there could be no colored sailors on Navy ships. FDR agreed. "It would be one thing if you could have Northern ships and Southern ships," he said, "but you can't do that." Instead, he suggested installing Negro bands on ships. "Because they're good at that."

Like many people who spent time with the charming President, the "race men" left the White House feeling they had accomplished something. Three weeks later, though, they were outraged when Press Secretary Stephen Early announced that there would be no change in the Navy's racial segregation policy. To make matters worse, he even made it sound as though they had agreed.

The Black leaders published a copy of the agenda they had taken to the meeting, and fired off a hot telegram to the President. Coining a phrase FDR had recently used against the French, they accused him of 'stabbing democracy in the back." FDR made a half-hearted public apology; he was sorry there had been such a misunderstanding on both sides.

The following spring, Randolph organized the Negro March on Washington Movement.[62] It was an intentionally drastic move to

> ". . . shake official Washington and the white industrialists and labor forces of America to the realization that Negroes mean business about getting their rights as American citizens under national defense."

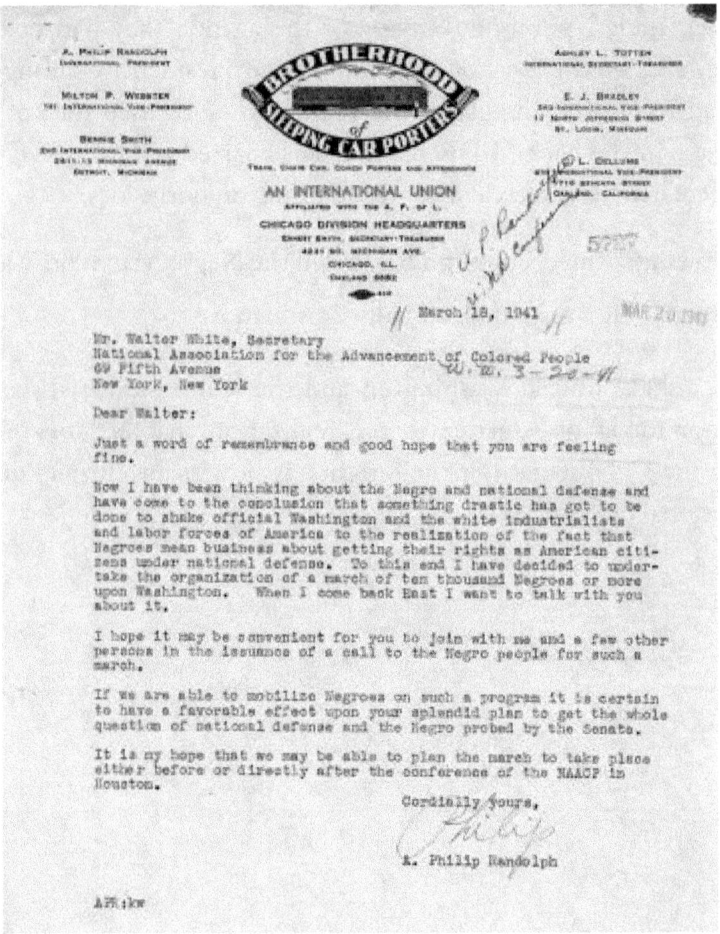

A. Philip Randolph
International President

Milton P. Webster

Bennie Smith

Ashley L. Totten

E. J. Bradley
St. Louis, Missouri

C. L. Dellums
Oakland, California

BROTHERHOOD of SLEEPING CAR PORTERS

AN INTERNATIONAL UNION

CHICAGO DIVISION HEADQUARTERS

Chicago, Ill.

March 18, 1941

Mr. Walter White, Secretary
National Association for the Advancement of Colored People
69 Fifth Avenue
New York, New York

Dear Walter:

Just a word of remembrance and good hope that you are feeling fine.

Now I have been thinking about the Negro and national defense and have come to the conclusion that something drastic has got to be done to shake official Washington and the white industrialists and labor forces of America to the realization of the fact that Negroes mean business about getting their rights as American citizens under national defense. To this end I have decided to undertake the organization of a march of ten thousand Negroes or more upon Washington. When I come back East I want to talk with you about it.

I hope it may be convenient for you to join with me and a few other persons in the issuance of a call to the Negro people for such a march.

If we are able to mobilize Negroes on such a program it is certain to have a favorable effect upon your splendid plan to get the whole question of national defense and the Negro probed by the Senate.

It is my hope that we may be able to plan the march to take place either before or directly after the conference of the NAACP in Houston.

Cordially yours,

A. Philip Randolph

APR:kw

". . . Negroes mean business about getting their rights as American citizens under national defense. . ." - A. Philip Randolph to Walter White, March 18, 1941

Unable to stop the march by any other means, FDR capitulated. On June 25th, he signed EO 8802 authorizing the Fair Employment Practices Committee (FEPC).

7 6/19/41

93

File

THE WHITE HOUSE
WASHINGTON

June 7, 1941

MEMO FOR MAC

On this Williston matter, we are accomplishing an enormous amount for negros in the Army and accomplishing a little in the defense industries. But we expect to accomplish more and on the whole progress has been relatively rapid. x25-Misc x25

Tell Williston that the President is much upset to hear (yesterday) that several negro organizations are planning to March on Washington on July first, their goal being 100,000 negoes and I can imagine nothing that will stir up race hatred and slow up progress more than a march of that kind and the best contribution Williston can make is to stop that march.

x391

F. D. R.

x335-Misc.

"I can imagine nothing that will stir up race hatred and slow up progress more than a march of that kind . . . stop that march." - FDR, June 7, 1941

After dictating the terms of the Executive Order, Randolph called off the march. In his memoir, **Off the Record with FDR, 1942-1945,** aide Bill Hassett divulged FDR's real attitude toward the "troublemakers."

"The Ethiopian in the wood pile"

Two important details form the backdrop for understanding how this period relates to the PCnet. First, FDR was embarrassed by the Negro leaders and appalled at the thought of the Negro March on Washington.

Second, on June 28, 1941, three days after he signed the FEPC into law, FDR signed Executive Order 8807, establishing the National Defense Research Committee (NDRC), headed by Vannevar Bush.

One of the primary responsibilities of the NDRC, an independent agency in the Executive Branch, was to do the footwork leading up to the American atomic bomb project. (It would require more imagination than research to discern the tenor of off-the-record discussions that may have taken place in the White House during those days.)

Sometime during this period, FDR reversed his position on Negroes in the Navy. By January 1942 he felt certain that the Navy could "invent something" for the colored sailors to do. He was determined to play a hands-on role in determining how and where the Black sailors would serve.

Two major events that relate to the PCnet coincided in the spring of 1942: 1) FDR authorized the Manhattan Project and 2) ground was broken at the Port Chicago Naval Ammunition Depot.

In March 1942, FDR insisted that he and Secretary Knox must be the ones to make decisions about where Negro sailors would be assigned. In April, the Navy announced that it would begin admitting Negroes to general ratings.

Two months later, in June, the first Negro enlistees arrived at Camp Robert Smalls, the Negro boot camp at the Great Lakes Naval Training Center near Chicago, Illinois. That same month, FDR authorized Bush's new more powerful agency, the Office of Scientific Research and Development (OSRD), and the Manhattan Project soon got underway.

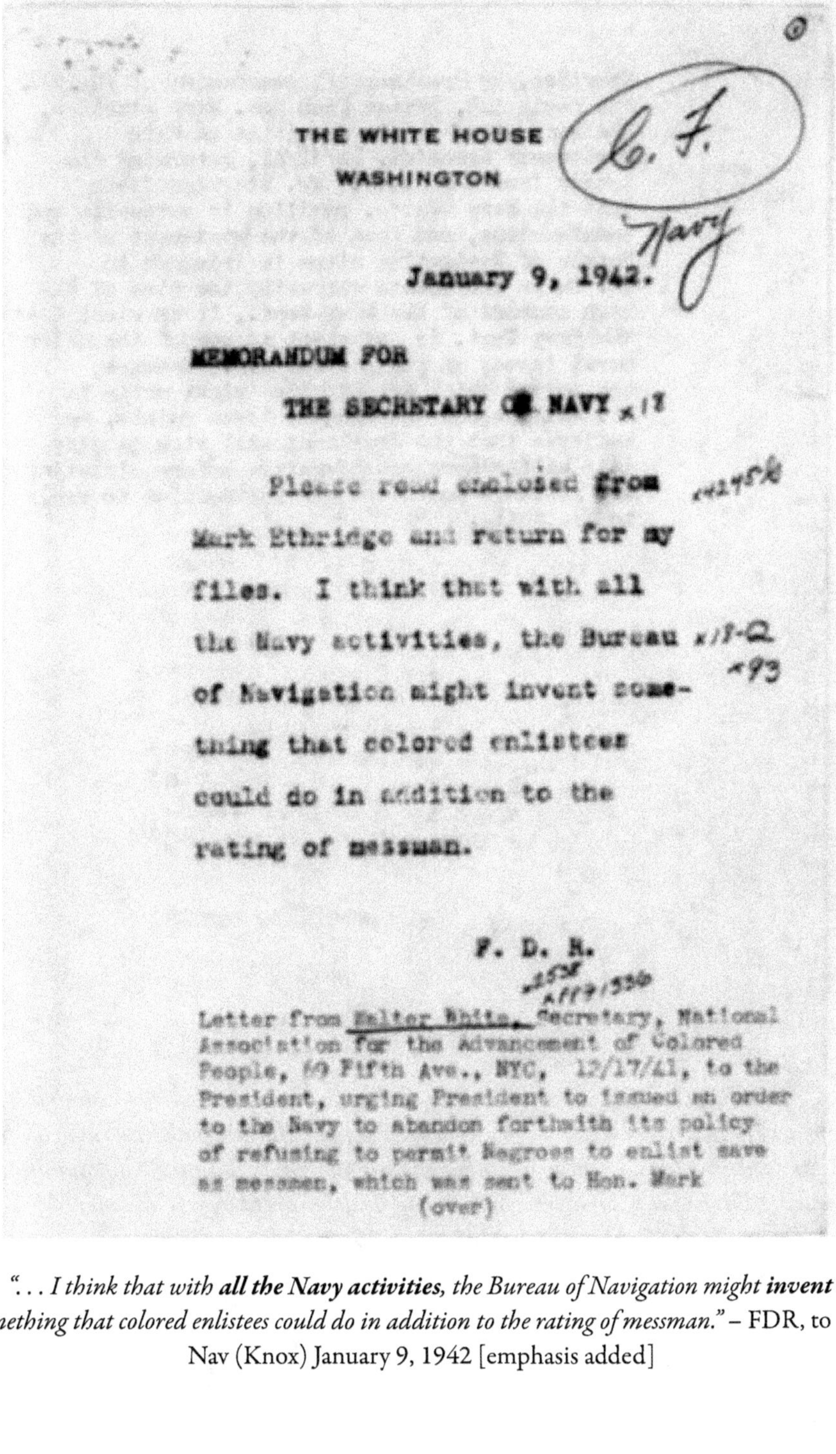
C. F.

Navy

THE WHITE HOUSE
WASHINGTON

January 9, 1942.

MEMORANDUM FOR

THE SECRETARY OF NAVY

Please read enclosed from Mark Ethridge and return for my files. I think that with all the Navy activities, the Bureau of Navigation might invent something that colored enlistees could do in addition to the rating of messman.

F. D. R.

Letter from Walter White, Secretary, National Association for the Advancement of Colored People, 69 Fifth Ave., NYC, 12/17/41, to the President, urging President to issued an order to the Navy to abandon forthwith its policy of refusing to permit Negroes to enlist save as messmen, which was sent to Hon. Mark
(over)

". . . I think that with ***all the Navy activities,*** *the Bureau of Navigation might* ***invent*** *something that colored enlistees could do in addition to the rating of messman."* – FDR, to Sec Nav (Knox) January 9, 1942 [emphasis added]

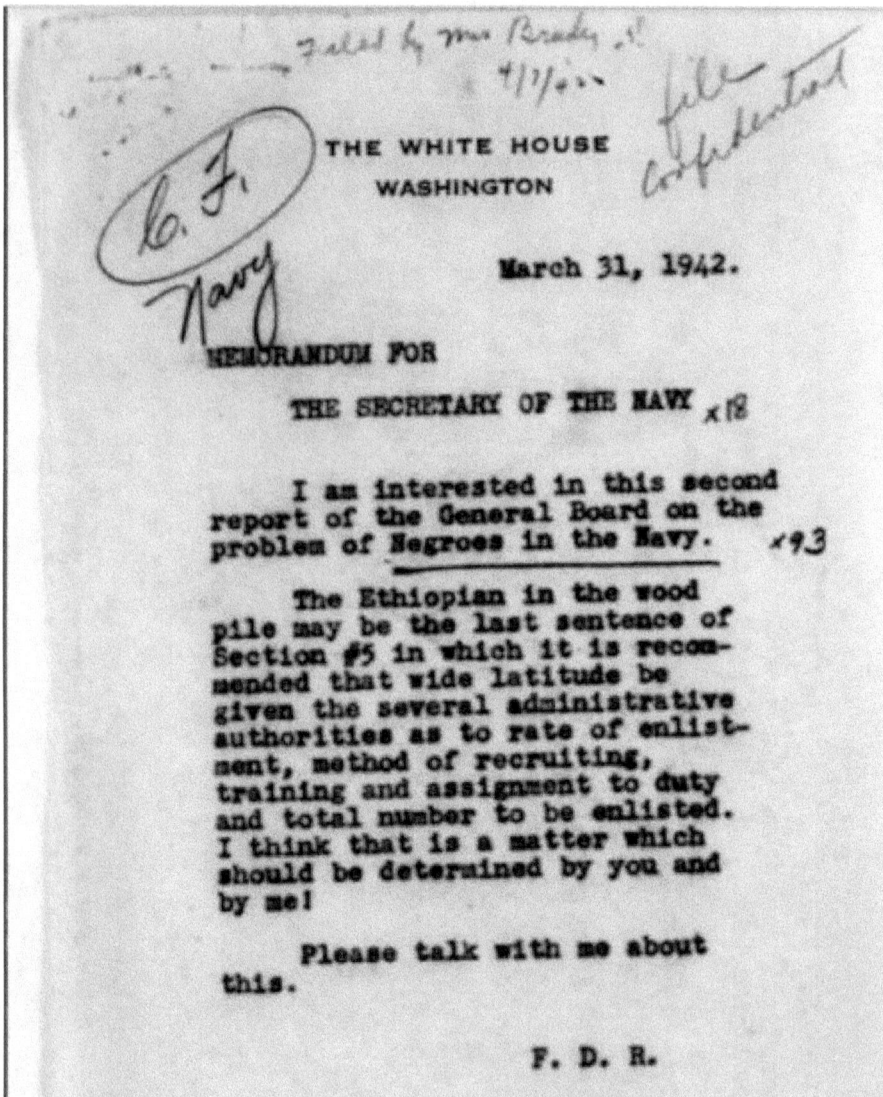

THE WHITE HOUSE
WASHINGTON

March 31, 1942.

MEMORANDUM FOR

THE SECRETARY OF THE NAVY

I am interested in this second report of the General Board on the problem of Negroes in the Navy.

The Ethiopian in the wood pile may be the last sentence of Section #5 in which it is recommended that wide latitude be given the several administrative authorities as to rate of enlistment, method of recruiting, training and assignment to duty and total number to be enlisted. I think that is a matter which should be determined by you and by me!

Please talk with me about this.

F. D. R.

“. . . *The Ethiopian in the wood pile may be the last sentence of Section #5 in which it is recommended that wide latitude be given the several administrative authorities as to the rate of enlistment,* ***method of recruiting, training and assignment to duty and total number to be enlisted. I think that is a matter which should be determined by you and me!*** *. . .*” - FDR, to Sec Nav (Knox) March 31, 1942 [emphasis added]

The Editor Says

THE NAVY AND THE NEGRO

THERE is nothing that we can see in the new policy of the Navy Department relative to the admittance of Negroes which is cause for rejoicing or dancing in the streets. To the sweeping protests of hundreds and thousands of citizens of both races against racial discrimination the Secretary of the Navy after months of consideration and reflection has answered by merely extending the area of segregation and giving this plan the official sanction of the government of the "greatest democracy on earth." The pity of it is that the Secretary of the Navy and the Administration in Washington appear not to have the slightest inclination to give real democracy a chance.

They might have provided for the training of small units of white and colored boys together as an experiment perhaps. Vast experiments are being conducted with machines, why not experiments with men? They might have announced that in those universities which now admit Negroes as students they would also be admitted to the naval officer training classes that are being conducted under the joint auspices of the universities and the Navy Department. Or they might have asked the trustees of Hampton to permit the department to train students for the Navy and decreed that men would be assigned to that institution without regard to race or color in order to have a practical and realistic test of racial attitudes among the youth of the country. In short, the Navy Department of the

At one time we were convinced that the heads of the War and Navy Departments lacked moral courage. We were mistaken. What they lack is vision.

WARREN LOGAN

AS OPPORTUNITY goes to press word comes of the death of Warren Logan, for forty-six years the treasurer of Tuskegee Institute. He was not as widely or well known as Booker T. Washington, the founder of Tuskegee, but it is not too much to say that without his guidance, advice and scrupulous care of the funds for which he as treasurer was responsible Tuskegee would never have attained its remarkable place in the educational life of America.

One is moved to speculate when time marks the passing of a man whose work and service reveal extraordinary talents and gifts. Particularly is this true of a colored man. What might he have been perhaps in the world of finance, or of industry, or of commerce, had it not been for the incident of color or the accident of race. In the case of Warren Logan such contemplation may be answered, that whatever he might have achieved in a world free from the cruel limitations which imprison capability within the high walls of prejudice, he could not have rendered greater service to his fellow man.

"There is nothing that we can see in the new policy of the Navy Department relative to the admittance of Negroes which is cause for rejoicing or dancing in the streets...Vast experiments are being conducted with machines. Why not with men?" – Editorial, E Elmer Anderson Carter - **Opportunity Magazine** [Urban League], May 1942

On July 15, 1942, FDR wrote a letter to the NAACP. He could not attend their annual convention, but took note of their slogan, "Victory is Vital to Minorities." To FDR, the reverse was also true. His words ran in bold headlines in newspapers across the country: "Minorities are Vital to Victory."

In February 1943, FDR bristled at the thought that he may have said something "in his sleep" about barring Negroes from the Navy. "Most decidedly" they were to be admitted; he could think of a thousand ways they could serve; and it would not require the intermingling of black and white sailors on ships.

CP CONFERENCE

ED OFFICERS

FDR TELLS NAACP "MINORITIES ARE VITAL TO VICTORY"

STATE NEGRO NYA OFFICE CLOSED

J. P. Bond Appointed To Washington Post

In September 1940, FDR opposed integration of the Navy. In April 1942, the Navy opened to Negroes. The first Black sailors were available to serve at Port Chicago when the naval ammunition depot opened in December.

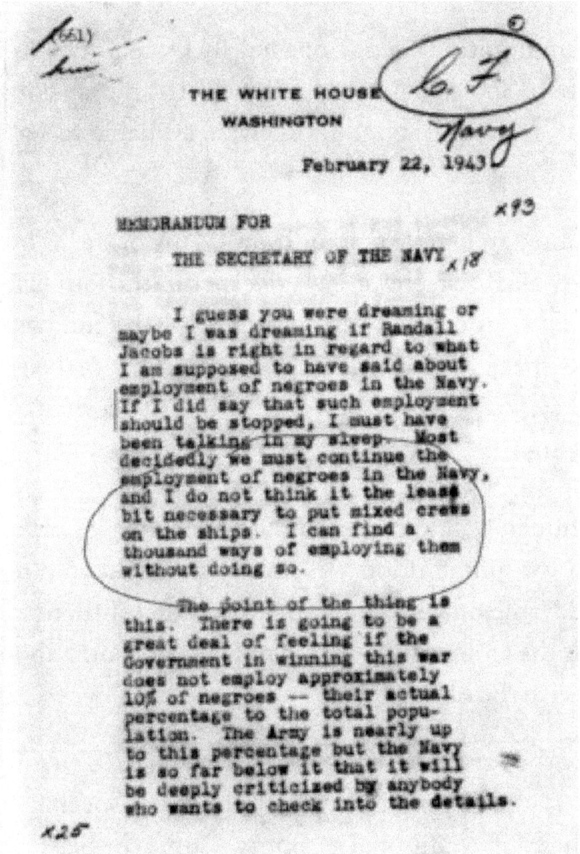

THE WHITE HOUSE
WASHINGTON

February 22, 1943

MEMORANDUM FOR

THE SECRETARY OF THE NAVY

I guess you were dreaming or maybe I was dreaming if Randall Jacobs is right in regard to what I am supposed to have said about employment of negroes in the Navy. If I did say that such employment should be stopped, I must have been talking in my sleep. Most decidedly we must continue the employment of negroes in the Navy, and I do not think it the least bit necessary to put mixed crews on the ships. I can find a thousand ways of employing them without doing so.

The point of the thing is this. There is going to be a great deal of feeling if the Government in winning this war does not employ approximately 10% of negroes -- their actual percentage to the total population. The Army is nearly up to this percentage but the Navy is so far below it that it will be deeply criticized by anybody who wants to check into the details.

"Most decidedly we must continue the employment of negroes (sic) in the Navy, and I do not think it the least bit necessary to put mixed crews on the ships. ***I can find a thousand ways of employing the without doing so."*** - FDR, to Sec Nav (Knox). February 22, 1943 [Emphasis added]

The Port Chicago Naval Ammunition Depot opened in December 1942. Graduates from Camp Robert Smalls (the segregated boot camp at Great Lakes Naval Training Center near Chicago, Illinois) were available to serve as stevedores at the base.

The sailors had not been trained to handle ammunition. Many of the White officers who supervised the all-Black crews lacked experience both with heavy explosives and with supervision of Negro personnel. (The following year, the Navy published a special manual for the Supervision of Negro Personnel.) By the fall of 1943, some of the sailors began to suspect they had "been "tricked" into something."

Safety hazards at the base, though well-known, went unaddressed.[63] Sailors were told on one hand that the ammunition was not armed and could not blow up; on the other hand, some officers reportedly joked that if there was an explosion, those handling the ammunition would not have to worry about it (since they would not survive the blast.)

As noted earlier, the cause of the explosion was never determined. (The COI conclusions will be discussed later.) Though the Court of Inquiry specifically stated that it found no one at fault, traditional accounts assume the blast was an accident caused by poor handling of ammunition by untrained personnel working under inexperienced officers who sometimes conducted races to see which team could load the ammunition first. Under such circumstances, an accident was foreseeable – and avoidable.

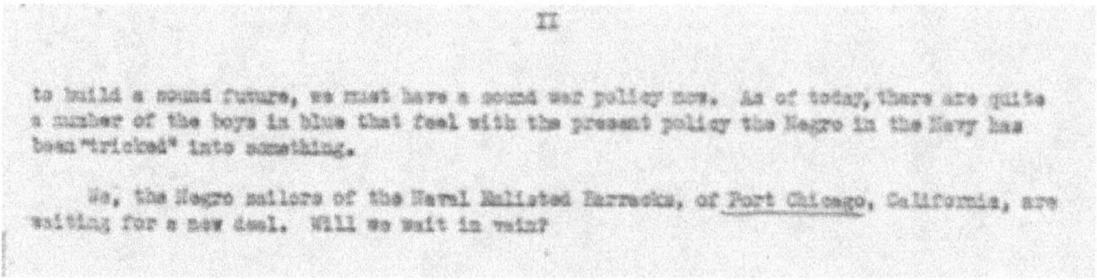

II

to build a sound future, we must have a sound war policy now. As of today, there are quite a number of the boys in blue that feel with the present policy the Negro in the Navy has been "tricked" into something.

We, the Negro sailors of the Naval Enlisted Barracks, of Port Chicago, California, are waiting for a new deal. Will we wait in vain?

"... As of today, there are quite a number of the boys in blue that feel with the present policy the Negro in the Navy has been "tricked" into something." - Negro sailors, Port Chicago Naval Ammunition Depot - Fall 1943

What the sailors could not have predicted was that the force of the blast would be of vital interest to key men involved in another secret race – the race to be the first nation to build an atomic bomb.

The PCnet emphasizes the fact that until they obtained the data from the Port Chicago explosion, Manhattan Project scientists and engineers lacked the information they needed to make reasonable predictions about the new, unprecedented weapon. As discussed later, the need for such information was known as early as November 1941 when Frank Jewett, head of the National Association of Scientists, advised Vannevar Bush that the data from the deadly Halifax explosion of 1917 would not be useful.

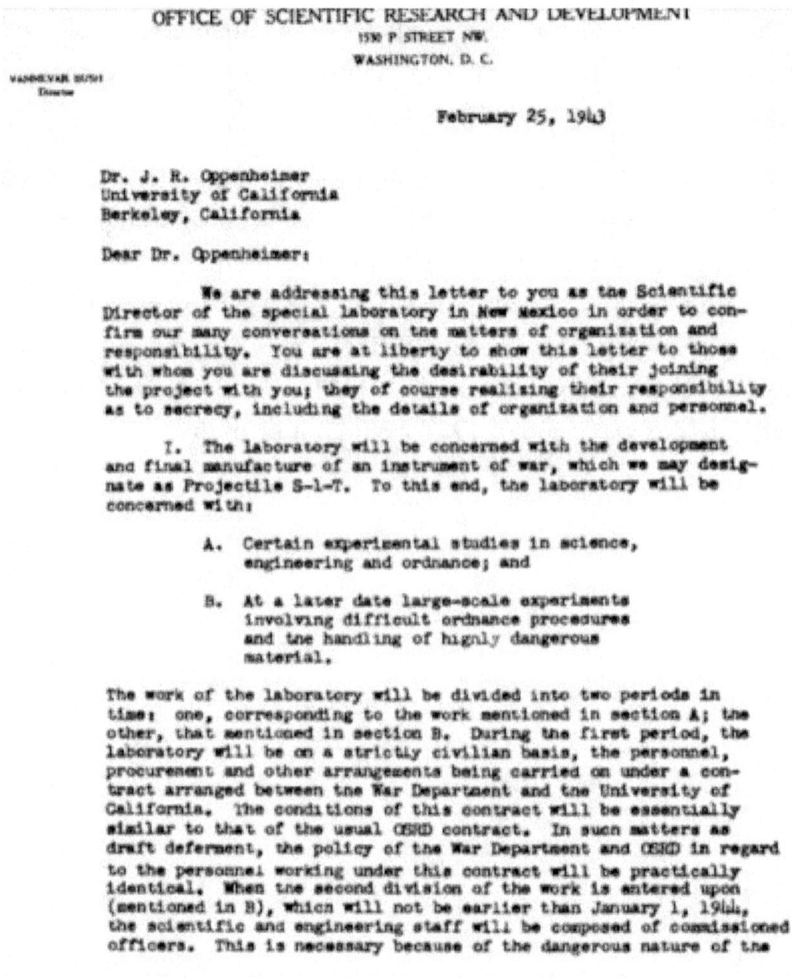

OFFICE OF SCIENTIFIC RESEARCH AND DEVELOPMENT
1530 P STREET NW.
WASHINGTON, D. C.

VANNEVAR BUSH
Director

February 25, 1943

Dr. J. R. Oppenheimer
University of California
Berkeley, California

Dear Dr. Oppenheimer:

We are addressing this letter to you as the Scientific Director of the special laboratory in New Mexico in order to confirm our many conversations on the matters of organization and responsibility. You are at liberty to show this letter to those with whom you are discussing the desirability of their joining the project with you; they of course realizing their responsibility as to secrecy, including the details of organization and personnel.

I. The laboratory will be concerned with the development and final manufacture of an instrument of war, which we may designate as Projectile S-1-T. To this end, the laboratory will be concerned with:

A. Certain experimental studies in science, engineering and ordnance; and

B. At a later date large-scale experiments involving difficult ordnance procedures and the handling of highly dangerous material.

The work of the laboratory will be divided into two periods in time: one, corresponding to the work mentioned in section A; the other, that mentioned in section B. During the first period, the laboratory will be on a strictly civilian basis, the personnel, procurement and other arrangements being carried on under a contract arranged between the War Department and the University of California. The conditions of this contract will be essentially similar to that of the usual OSRD contract. In such matters as draft deferment, the policy of the War Department and OSRD in regard to the personnel working under this contract will be practically identical. When the second division of the work is entered upon (mentioned in B), which will not be earlier than January 1, 1944, the scientific and engineering staff will be composed of commissioned officers. This is necessary because of the dangerous nature of the

Research and development at the Los Alamos Laboratory would ultimately involve large-scale experiments. - Bush to J. Robert Oppenheimer, February 25, 1943

In a February 1943 letter to Oppenheimer, Bush laid out the general responsibilities of the Los Alamos laboratory, which would include, in the latter stages, large-scale experiments.

Due to the hazardous nature of the work involved, the second division of the work, the scientific and engineering, would be led by commissioned officers.

William S. Parsons would be the officer in charge of the ordnance experiments. He was recruited in May 1943 and officially joined the Project in June. However, given his credentials and background, as well as his connections to various parties connected to the Project and to uranium research (including Ross Gunn, Phil Abelson, and Vannevar Bush), it is reasonable to wonder how much Parsons knew about the Project prior to that time and whether or not he had any input in the planning stages.

In **Target Hiroshima,** Christman says he did not.

> "Even though Deak Parsons had been V. Bush's chief conduit for the proximity fuze program, then the exemplar of a successful secret weapon, he did not have access to his civilian boss's even bigger secret: the high-priority work then under way to build an atomic bomb."

Most sources identify Parsons as "the Navy captain who armed the *Little Boy* aboard the Enola Gay." Christman is the rare historian to both recognize Parsons' influential role in the creation of the Project and note the power the Navy captain had as a result of his connection to Bush.

But, as noted earlier, Christman also downplayed Parsons' visit to Port Chicago. One reason he wrote Parsons' biography was to counter the PCnet. He may have been more convincing, though, if he had explored the relationship between Parsons and his two colleagues who sat on the original Uranium Committee, Merle Tuve and Ross Gunn.

With respect to both personality and power, If there was a person who could have pulled off a test of an atomic device at Port Chicago, it was William S. Parsons.

Necessary ruthlessness

Parsons was influential. He was driven. As head of the ordnance division of the Manhattan Project, the Experimental Officer had a two-part goal: to create the perfect bomb and perfect its delivery. As the weaponeer on the Enola Gay, he accomplished that mission, and found the delivery "awesome."

Some sources paint him as a sympathetic person who hated having to obey the order to drop the bomb on Hiroshima. In **Day of the Bomb: Countdown to Hiroshima**, Dan Kurzman takes this sentiment to the extreme, even going so far as to describe Parsons' "dream "for the near future," as reflected in his support for the Atoms for Peace program . . .

> ". . . De-salting sea water for arid countries so farming could be done and starving people fed, a sharing of medical knowledge in the battle against such diseases as malaria; educational and cultural exchanges between all countries, and other ways to live peacefully in the world. If all these things came about, there would be no more wars. But until such a time came, the United States must keep defensively as strong as possible. It would be dangerous to have it any other way."[64]

According to Kurzman, Parsons told his wife, Martha, that August 6, 1945 was "the happiest day of his life" – but that was only "because it was no longer necessary to invade Japan."

And Parsons apparently convinced Tuve that it hurt to drop the bomb on Japan, that he had preferred a demonstration. But "orders were orders".

This source is one example of others that make similar assertions about Parsons' character, generally based on interviews with colleagues or family members. Such eulogies are not entirely surprising. Nor are they necessarily accurate. What makes them especially revealing is how they contrast with public statements by Parsons himself, and by people like General Thomas Farrell, one of the three "Tinian Joint Chiefs" (along with Parsons and

Admiral William R. Purnell). In the interview moments before Parsons described the perfect delivery of the weapon, Farrell expressed gratitude that 'we were given" the bomb, and prayed that it would only ever be used by good guys, for the good. (Meaning, presumably, not as a weapon of mass destruction.)[65]

Similar dissonance is apparent in contradictory statements made by President Truman, who crowed about the great achievement on one hand but in another instance expressed regret at having to 'treat a beast like a beast.' And this conflicting attitude is inherent in the MAD policy of mutually assured deterrence, which proclaims how wonderful it is that the terrible weapon has prevented another worldwide war.

That is, the bomb *seems* to have prevented a third world war. So far. In the first place, no one has proved – and no one *can* prove—that cause/effect relationship. And no one can predict that there will never be another war. If that happens, then at best the bomb would have delayed another war. What we can be certain about is the one thing we fear: nuclear weapons make nuclear war possible. If not inevitable.

As for a demonstration of the bomb, what Parsons said in writing is likely to more accurately reflect his true feelings.

In a September 1944 letter to Oppenheimer, he insisted that "Ruthless, brutal people must band together to force the *Fat Man* components to dovetail in time and space . . ." This fearless band of leaders, he said, "must feel that they have a mandate to circumvent or crush opposition above and below, animate and inanimate — even nuclear!"[66]

Christman further describes Parsons' concerns about not completing the bomb project.

> Parsons concluded that the mission could also be jeopardized by a weakening resolve to use the final weapon. He wrote Groves, "Some *tender souls* are appalled at the idea of the horrible destruction which the bomb might wreak in battle delivery. I

> believe these imaginations magnify the horror beyond what they read about in thousand-bomber raids." Parsons felt it would be but a short leap from "this loose reasoning to the expressed or unexpressed hope that 'We may never have to use this weapon in battle.'" Once that happened, in his view, the drive necessary to complete the project would be lost.[67] [Emphasis added]

Parsons spoke of “tender souls” not wanting to use the bomb. FDR used the term “timid souls” to describe those who would oppose a “bold strike.” Perhaps the language is just a coincidence.

Parsons was concerned that the hope of never having to use the weapon in battle could cause the Project to lose momentum, in which case the culmination of work on what he called “our bomb” would be a laboratory experiment. It is not clear precisely how he knew, but Parsons was aware that some discussion to that effect had recently taken place in high places.

That same month, FDR had raised the idea of a demonstration with Bush, who speculated that the President was speaking under the influence of physicist Niels Bohr. The esteemed scientist had met with both Churchill and FDR, pressing them to consider international control of the bomb.

On the other hand, September may have been the first time Bush spoke with FDR about the bomb since the President’s return from his trip to the West Coast and to Hawaii (that is, since the Port Chicago explosion). According to Bush, that September conversation was the only time FDR mentioned using the bomb in a demonstration.[68]

The idea of the atomic bomb as a means of convincing men to avoid war was established as early as 1914 when H. G. Wells published his influential novel, **The World Set Free**. FDR envisioned the United Nations as a means of enforcing international peace. In **The Age of Hiroshima**, Craig suggests that FDR may have seen the bomb as an essential means to securing Stalin’s cooperation with his plans for the UN.

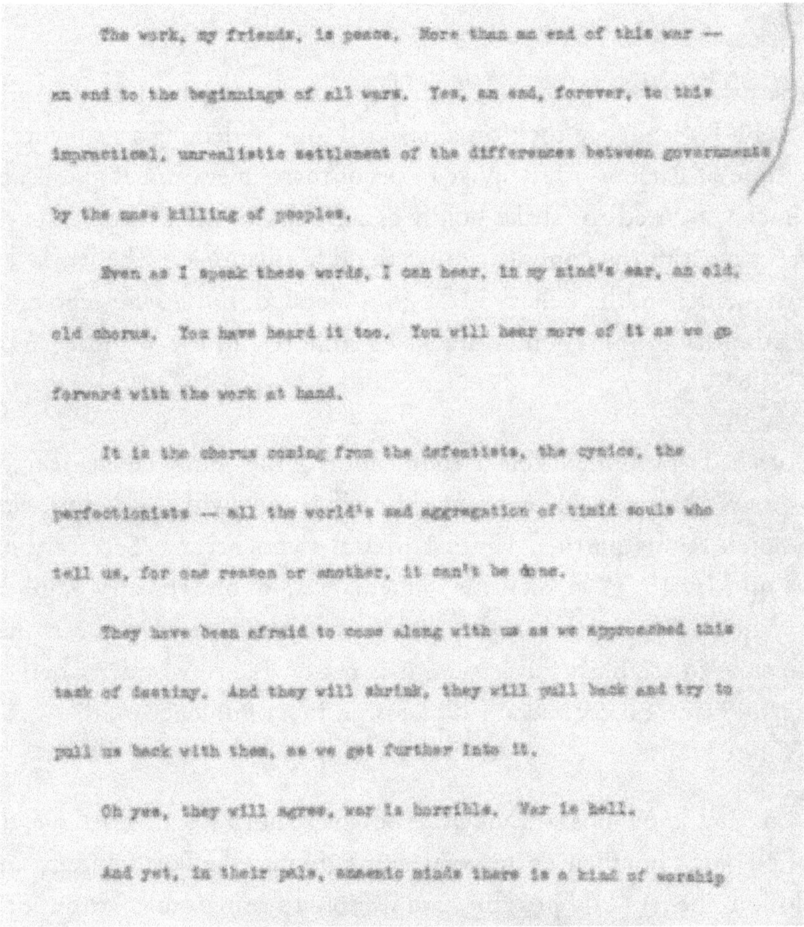

The work, my friends, is peace. More than an end of this war -- an end to the beginnings of all wars. Yes, an end, forever, to this impractical, unrealistic settlement of the differences between governments by the mass killing of peoples.

Even as I speak these words, I can hear, in my mind's ear, an old, old chorus. You have heard it too. You will hear more of it as we go forward with the work at hand.

It is the chorus coming from the defeatists, the cynics, the perfectionists -- all the world's sad aggregation of timid souls who tell us, for one reason or another, it can't be done.

They have been afraid to come along with us as we approached this task of destiny. And they will shrink, they will pull back and try to pull us back with them, as we get further into it.

Oh yes, they will agree, war is horrible. War is hell.

And yet, in their pale, anaemic minds there is a kind of worship

". . . The work, my friends, is peace. More than an end to this war – ***an end to the beginning of all wars.*** *Yes,* ***an end, forever, to the impractical, unrealistic settlement of the differences between governments by the mass killing of peoples.*** *Even as I speak these words, I can hear, in my mind's ear, an old, old chorus. You have heard it too. You will hear more of it as we go forward with the work at hand.*

It is the chorus coming from the defeatists, the cynics, the perfectionists – all the world's sad aggregation of ***timid souls*** *who tell us, for one reason or another, it can't be done . . ."*

- FDR, draft of final undelivered speech, April 13, 1945 *[emphasis added]*

In the draft of his final, undelivered speech for Jefferson Day (April 13, 1945) FDR's language took on a hawkish tone, reflecting sentiments more like those of Parsons. FDR spoke of proof that Americans, Republican and Democrat, wanted to "strike boldly against the threat of war." His recent travels gave him the confidence that the other nations of the world would feel the same, and he believed "the thin-blooded, *timid souls* who are now in a minority in our country are also a minority in the world." [emphasis added][69]

Parsons had a prominent role in the creation of the bomb. He made frequent trips to Washington. He enjoyed close affiliations with Bush and with his occasional tennis partner, James Forrestal (who became Secretary of the Navy on May 19, 1944, after his immediate superior Secretary Frank Knox died from a heart attack.). On the basis of these facts, while it may be impossible to establish for certain, it is reasonable to wonder whether the powerful Naval officer had any direct or indirect influence on the "Sailor in the White House."

Parsons' mark on postwar nuclear policy is quite clear. His role in the creation of the bomb gives the greatest insight into the Port Chicago nuclear explosion theory. His possible connection to and influence on FDR is worthy of further research.

"The source of his power"

Admiral William S. Parsons was one of the most powerful men ever hidden in history. He was instrumental in developing each of the three military technologies that were key to the victory in World War II – radar, the proximity fuze, and the atomic bomb. Parsons would have been the person in charge of a nuclear test at Port Chicago. But how would he have done it, and why?

As we will see, Parsons had the necessary power, opportunity and means to conduct the secret operation. He had ultimate control of the atomic weapons in production. He directed the testing of the *Thin Man* and *Fat Man* dummy (non-explosive) test models. He hired the men who worked on them and conducted the research. He had used the Port Chicago Ammunition Depot for storage and transshipment of parts in the past, so was familiar with the facility. He had a perfect cover.

Additionally, Parsons had a strong propensity for conducting tests and even had a compelling motive to test a weapon in a full-scale, combat-like setting: Parsons was determined to deliver the perfect weapon. For that, the Experimental Officer whose career centered on testing new technologies would have no doubt assumed at least one test would be required.

Parsons was born in Chicago, Illinois on November 26, 1901, but his family moved to New Mexico when he was 8 years old. In 1918, he went to the Naval Academy at Annapolis, and graduated in 1922. After a five-year tour of duty, he returned to Annapolis to do graduate studies in ordnance engineering at the Naval Postgraduate School. From there, he went on to the Naval Proving Ground in Dahlgren, Virginia, where he studied ballistics.

In 1933, after another tour, he became liaison officer between the Bureau of Ordnance and the Naval Research Laboratory (NRL) in Washington, DC. There, he developed a proposal for use of radar by the Navy. After some bureaucratic challenges, his idea gained the attention of Admiral Ernest

King, who not only supported radar but continued to be one of Parsons' many powerful contacts throughout his career.

In 1939, Parsons was back at Dahlgren, where his keen interest in science and his visionary role in developing radar for military use made him an ideal candidate to lead the research and development of the proximity fuze. In that role, as in others, he used his notable ability to speak the language of both science and the military. Through his work on the proximity fuze, Parsons became Special Assistant to Vannevar Bush, head of the Office of Scientific Research and Development (OSRD), which in turn led to his position with the Manhattan Project.

In May 1943, Parsons, promoted to Captain, joined the Manhattan Project as head of ordnance, the man responsible for the creation and delivery of the atomic bomb. In August 1944, he became one of two Associate Directors of the Los Alamos Laboratory. (The other Associate Director was Nobel-prize winner Enrico Fermi, the physicist who led the December 1942 experiment at the University of Chicago that demonstrated the possibility of a nuclear chain reaction.)

In 1946, Parsons served as Technical Director over *Operation Crossroads*, the grand-scale tests of the atomic bomb against naval ships in the Bikini Islands. And, before his sudden death from a heart attack in December 1953, Parsons became a Rear Admiral, the first "atomic admiral."

Historical accounts attribute his fatal heart attack to his distress upon learning that Oppenheimer's security clearance was to be revoked – as if an otherwise healthy and relatively young military officer was apt to succumb to death as a result of someone else's bad news. To make the ridiculous idea sublime, the decision had not yet been made; in fact, Oppenheimer himself insisted on the hearing, which had yet to be scheduled.

No matter how close the relationship was between Parsons and Oppenheimer, however close their families were, there simply must be more to the story. Instead of probing, though, sources simply repeat the charming tale that seems to portray the Rear Admiral as a loyal friend.

The explanation is especially dubious when we realize that Oppenheimer opposed at least two of Parsons' projects: underwater delivery of the bomb (i.e., an atomic torpedo) in 1943 and *Operation Crossroads* in 1946. We will revisit this topic later. Some background will put this information in perspective.

The level of Parsons' power is inverse to the degree of his fame. There is reason to believe his power and his obscurity are related: not only has he been hidden in history, so have his most notable accomplishments. He believed ". . . there is no limit to the good a man can do if he doesn't care who gets the credit."[70]

Sources attribute this sentiment to Parsons' humility. But the benefit of tight lips also applies to secret missions. One key to the success of the Manhattan Project was General Groves' insistence on compartmentalization: information was shared on a strict 'need-to-know' basis.

Like history, biography is narrative; certain facts are selected and highlighted, others omitted. A particular story is told in a particular way to convey particular ideas, while downplaying or concealing others. Despite his remarkable career, Parsons is best known as "the Navy Captain who armed the *Little Boy* bomb in midair."

Parsons was an observer on the plane that flew over the Trinity test of July 16, 1945. He famously served as weaponeer aboard the *Enola Gay* for the attack on Hiroshima. Though he did not witness the attack on Nagasaki, he was part of the research team that studied the effects of that bombing. Parsons was Technical Director for *Operation Crossroads*. He was the first person to suggest establishing a proving ground in the continental U.S., a suggestion that led to creation of the Nevada Proving Ground.

It is an understatement to say that Parsons was very well-connected. In May 1943, after meeting with Groves to talk about the new job, Parsons visited Bush and Rear Admiral H. P." Spike" Blandy, the two men who were "the source of his power in the fuze program," according to biographer Al Christman. (Blandy was the admiral in charge of *Operation Crossroads*.)

During the war, Admiral H.R. Purnell, the Navy representative on the Military Policy Committee that advised Bush on the Manhattan Project, decided to keep Parsons on the official staff of Admiral Ernest King. As Christman points,

> "...Purnell and Parsons quickly recognized that there could be long-range advantages in keeping Parsons' name on the rolls of Admiral King's staff. For one thing, it provided security cover for Parsons, a plausible explanation of his whereabouts. When well-meaning friends saw him in Washington and wanted to know what he was doing, the answer was simple: "I'm on Admiral King's staff."[71]

The cover that allowed Parsons to come and go as he pleased in Washington would apply as well to other places. Even prior to the war, for instance, he had access to the Port Chicago Naval Ammunition Magazine, where he had stored material during the tests of the proximity fuze. Under his leadership, the ordnance division of the Manhattan Project had easy access to the military base, which was, of course, off limits to the press and the public.

Parsons traveled to Washington at least once a month, if for no other reason than to meet with Groves. Whatever else he accomplished on those trips, it would require further research to determine whether or not the prominent Naval officer ever had a personal encounter with President Roosevelt. The possibility is real, the implications worth investigating. After all, Parsons – the singular ordnance expert who was read into the secret – had direct access not only Bush but to several of the men who set policy for the atomic bomb project. They trusted him to represent their interests. "More important," Christman explained,

> ". . . the phantom position gave him the authority to originate high-priority requests and purchases out of King's office without disclosing the Los Alamos connection. The ComlnCh [Commander in Chief] position and the backing of Admiral Purnell added to Parsons' unique powers within Project Y. In

> short, Parsons went into the bomb project with a direct line to the summit of the U.S. Navy. He wore this power, as well as his influence with Vannevar Bush of the scientific community and Spike Blandy of BuOrd [Bureau of Ordnance], much like a concealed weapon. He knew it was there. Groves and Oppenheimer knew it was there. Project leaders at Los Alamos sensed its presence. Yet possessing the weapon, Parsons brandished or used it only when essential.[72]

Even before he began working with the Project, Parsons had access to whatever resources he needed, whether money, men or material. While working on the proximity fuze, for example, he had commandeered a cruiser, the USS *Cleveland*, for use in field tests, and requisitioned six radio-controlled aircraft to serve as pilotless drone targets. (He only got four.)

This is not the portrait of a man whose claim to fame is the one act—however glamorous, heroic and unique—of arming the *Little Boy* bomb in midair. With that minimization of his role, which contributes to his obscurity, Parsons' cover was solid and permanent. If ever anyone could have pulled off a clandestine test of a nuclear device at the remote naval ammunition magazine, Captain William S. Parsons was that man.

The quest for a nuclear torpedo

During the war, Parsons investigated the possibility of a nuclear torpedo.[73]

Several points are evident without engaging in technical details, which are beyond the scope of this book. First, it is clear that Parsons was in pursuit of underwater delivery of a nuclear weapon in the fall of 1943.

It is clear that he engaged the services of John von Neumann and other experts. Experiments were done on a very small scale at first, but according to David Hawkins, author of **Manhattan District History – Project Y, the Los Alamos Project: Inception until August 1945**, the experiments were "scaled up to explosions of the order of magnitude of interest to the Laboratory."[74]

It is evident that the Port Chicago explosion was on the "order of magnitude" of interest to the ordnance division.

In the fall of 1943, Parsons met with mathematician John von Neumann at Los Alamos. Their apparent focus was on the study of implosion as a means of detonating an atomic bomb, though implosion was not yet a priority for the Project. von Neumann, an expert on underwater explosions, worked at Princeton's Institute for Advanced Studies, where he was a colleague of Albert Einstein. On at least one occasion in the summer of 1943, Einstein consulted with von Neumann regarding his work on torpedo design for the Navy.

In November 1943, Parsons was pursuing underwater delivery. Oppenheimer, expressing concern that the method did not appear to be promising, suggested that Parsons should focus on other methods. There seems to be no indication, however, that Parsons discontinued the work altogether.

In his Aug. 5, 1997 article for the **San Francisco Chronicle**, titled "*It's History - The Quest for an Atomic Torpedo*", historian Barton J. Bernstein noted that Oppenheimer's objection

> "... was not an order to quit, but rather one to reduce hopes for a nuclear torpedo.
>
> Such counsel probably did not kill the project, but soon other problems emerged. Physicists concluded that a gun mechanism assembled in a plutonium weapon would be so unstable that it might blow up American forces before they could use it. And estimates indicated that any likely nuclear weapon would be too much too large and heavy for delivery from a torpedo tube."
>
> The hopes and early planning for a nuclear torpedo collapsed by early or mid-1944. . ."[75]

Given his adamant posture on testing of radar and the proximity fuze, it would have been uncharacteristic for him to yield to Oppenheimer's discouragement. As an experimental engineer, Parsons enjoyed finding solutions to difficult problems. To Bernstein, the quest for an atomic torpedo

> ". . . is a useful reminder of the secret World War II interest in developing nuclear weapons besides the airplane-dropped bomb, and suggests the desire by at least one Navy man to give his own military service a prominent role in using atomic weapons in World War II."[76]

A *Crossroads* connection

The full context for *Operation Crossroads*, the 1946 tests of atomic bombs against naval ships, includes the ongoing rivalry between the Army and the Navy, which became a three-way contest as the Air Force demonstrated its prowess during World War II. (The Air Force was still part of the Army during World War II.)

In 1925, shortly after Parsons graduated from the Naval Academy at Annapolis, another prominent "Navy" man, Assistant Navy Secretary Franklin D. Roosevelt, followed the court martial of Billy Mitchell, the disgraced Army Air Force General who defied his superiors in his bodacious effort to prove the superiority of airplanes over ships. FDR's boss at that time, Navy Secretary Josephus Daniels, defended the Navy, declaring that

> "I would be glad to stand bareheaded on the deck or at the wheel of any battleship while Mitchell tried to take a crack at me from the air. If he ever tries to aim bombs on the decks of naval vessels, he will be blown to atoms long before he gets close enough to drop salt on the tail of the Navy."[77]

FDR was more pragmatic. At the beginning of hostilities in 1939, he issued an appeal for belligerent nations to refrain from attacks on civilians. By his death in 1945, however, he was presiding over the "total warfare" that included firebombing of cities in Europe and Japan.

In September 1945, twenty eventful years after the Billy Mitchell trial, Parsons met with Admiral H.R. Purnell to discuss proposals for tests of atomic bombs against Naval ships.

Like many aspects of nuclear history, *Operation Crossroads* has one accepted origin story and a lesser-known reality. In the official account, the proposals were sparked by Senator Brien McMahon (D-CT) and Admiral Ernest King. In actuality, Parsons himself had prepared a paper on the subject in July 1944.[78]

The ongoing rivalry between the Army and the Navy was part of the impetus for the tests. As Assistant Secretary of the Navy during the Billy Mitchell scandal, FDR had considered it unlikely that airplanes could ever prevail against Naval ships. As wartime President, though, he came to realize the importance of airplanes. Whether or not he ever discussed the transition from airplanes to the atomic bomb with his military science advisor, Bush – or anyone else – would be a subject for further research;

It is certain, however, that in July 1944 when he wrote his pilot study, "*Comments on Planning for Army and Navy Research*," Parsons was influenced by the 1922 tests of airplanes against ships. The up-and-coming naval officer was aware of both the inter-service rivalry and the importance of new military technologies. Then in the fall of 1945, Purnell tapped him to head up a study on the feasibility of testing the bomb against ships. Christman says,

> "Parsons welcomed the task; this was a subject on which he was uniquely qualified. *As part of his wartime work he had studied the theoretical effects of atomic weapons on naval vessels* and had even pursued the idea of developing a torpedo to deliver an underwater nuclear explosion against ships[fn3]. These wartime studies were unfinished business, and Parsons did not like loose ends."[79] [emphasis added]

Throughout his career, Parsons was a forward-thinking ordnance engineer who foresaw and prepared for the next set of necessary experiments. During the war emergency, he had been able to do things that could not be done in peacetime. As the war drew to a close, he became concerned about bureaucratic "cognizance" and the territorial infighting that hampered creativity and progress. *Operation Crossroads* would give him a chance to go to the next level.

The progression of scientific research, theoretical and experimental, is seen again in the preparation for *Operation Crossroads*. In April 1946, Maurice Shapiro and von Neumann co-authored a study on "the principal phenomena which are expected to be of importance in the underwater explosion of a nuclear bomb."

This study is comparable to one in the appendix of Parsons' Third Preliminary Report on the Port Chicago explosion. The July 31st report, "Effects of the Tidal Wave in the Port Chicago Explosion," was prepared by Shapiro, who accompanied Parsons to Port Chicago in July 1944. The study compares the effects of the Port Chicago blast to those predicted by theoretical research.

In other post-war positions, as Assistant Chief of Special Weapons and as Rear Admiral, Parsons continued to push for nuclear testing. It was his belief that Americans were too afraid of radioactivity. To him, their fear was irrational and counterproductive. The need was so great that the government should be able to test nuclear weapons whenever and wherever it was necessary; Americans should welcome the tests within 100 miles of their homes.

Given his many accomplishments, as well as his connections, Parsons should be a well-known historical figure. From the perspective of the PCnet, though, there is a circular explanation for his obscurity.

The postwar campaign to sell the bomb to the American public stressed the glamorous scientific aspects. National security was the presumed reason why Project Director Leslie Groves made the decision to minimize references to ordnance and engineering. But if there was a nuclear test at Port Chicago, one way to maintain the secret was to downplay Parsons' key role in the research and development of the weapon.

Parsons at Port Chicago

When the Court of Inquiry (COI) found that the cause of the explosion could not be determined, it meant that there was no need for further investigation. That would have been a convenient ruling for someone who had something to hide – which makes it especially noteworthy that Capt. James R. Crenshaw, Parsons' brother-in-law, was one of the three men on the COI. To Vogel, their relationship posed a conflict of interest with regard to the mutiny trial. In **The Color of War**, James Campbell indicates that it was Parsons who arranged for his brother-in-law to be on the Court.

Crenshaw was with Parsons when he visited Port Chicago on July 20th. From his conversation with Admiral Wright, it is evident that Parsons was concerned not only with "the effects of the detonation" but with the makeup of the Court. This makes little sense if his interest was merely in the data on the effects of an *accidental* explosion (which Parsons referred to as "the detonation.")

Campbell—who may have underestimated Parsons' power and autonomy—says he was sent to Port Chicago, but does not indicate by whom or why. Christman, on the other hand, says Parsons went to Port Chicago on his own, as the senior naval officer at Los Alamos. He did not have to go, Christman said, but wanted to see for himself. The explosion would give him a chance to see the effects of the largest explosion in the U.S. Because neither writer provides support for their conclusions, these otherwise reasonable explanations are tantamount to speculation, and cannot be relied on as historical fact.

Parsons himself explained his purpose for going to Port Chicago in his preliminary memo. He emphasized the purpose of his mission in a brief conversation with Admiral Wright. He was there "to obtain data on effect rather than the cause of the detonation."

There are several points to consider here. First, there is a question of interpretation. Just as in the question of what he meant by "Port Chicago was designed for large explosions," his choice of words matters here.

For one thing, he was on a mission. From the letterhead, it appears to be an internal Navy mission; there is no indication that he was head of ordnance for the Manhattan Project. Given the full context, though, it is clear that Parsons was the person making the decisions. No one sent him. He had other reasons for being there. Other agencies were there to study the explosion, but Parsons' unique interests could not have been met by reading even their most extensive reports.

The "detonation"

Parsons was interested in the effect, not the *cause* of the explosion – or, as he referred to it, "the detonation". He may have already learned from the COI report (or from his brother-in-law, a member of that board) that the cause could not be determined. Without knowing the cause, it is not clear why he would have been confident that data on the effects of a conventional explosion would be useful in the development of an atomic weapon.

He could not have assumed the usefulness of an accidental explosion caused by mishandling of conventional ammunition on the pier and on the ship. After all, the Port Chicago data was of critical importance *because* data from the Halifax explosion of 1917 was skewed due to other factors. This is not mere speculation. In a postwar speech to the Naval War College, Parsons explained how the reinforced design of Port Chicago benefited their study of the effects. "With some luck," he said, "this protection was effective in preventing any secondary fires."[80]

Secondary? Secondary to what? The *primary* results of "the detonation"? That is one of the first (previously unasked) research questions that needs to be addressed. Then, there is the term "detonation" to consider. By definition,

> "A detonation is an explosion, usually done on purpose. Nuclear weapons cause massive detonations, and cities sometimes plan a

> careful detonation of an abandoned building in order to make room for a new structure.
>
> Detonation can be two things: the actual moment someone presses the button that creates the explosion, and then there's the explosion itself. Its root word is the Latin *detonare*, which means "thunder down," and if you've ever heard a detonation you understand that phrase well."[81]

An *explosion* may be intentional or accidental; a *detonation* is typically an event that is triggered by intention. (One exception, of course, is predetonation, caused by spontaneous fission, which was what led to the shelving of the *Thin Man* on July 17, 1944. We'll come back to that.)

Parsons was on a mission: he had an intention. It was to gather data that was essential to the top-priority project, the urgent creation of the perfect atomic bomb. Not just any data from any explosion would do. For one thing, they were not certain what to expect from an atomic explosion; they were certain that the effects would differ from those resulting from a conventional blast.

By one interpretation, Parsons was dissatisfied with the results of the Port Chicago detonation. In a September 1944 letter from Parsons to Groves, via Oppenheimer, Parsons used the phrase "my observation of the Port Chicago explosion . . ." Apparently, based on what he had *seen*, Parsons concluded that foreign observers would be unimpressed with a "desert shot." They would find even the crater disappointing.[82]

In comparing "a desert shot" (i.e., the Trinity test, a demonstration shot) to the Port Chicago explosion, Parsons apparently considered the Port Chicago explosion unimpressive. The blast that destroyed the pier, killed 320 people and destroyed two ships did not meet his expectations. But on what basis would the effects of an *accidental* explosion have fallen short of his expectations? What expectations would he have had?

But, as we will see, there is another, more chilling interpretation that can be supported from statements Parsons made in the letter.

Was the *Thin Man* a disappointing fizzle?

The *Thin Man* plutonium gun-assembly bomb was the weapon that had been the main focus of the ordnance effort for over a year. Scientists had predicted that plutonium would predetonate in a gun-assembly weapon, producing an explosion that was less than optimum—a fizzle. If Parsons found the Port Chicago explosion disappointing, maybe that's because it was a fizzle.

An atomic fizzle is not like the sputter of a flawed firecracker. Historian Richard Rhodes, in his highly acclaimed book **The Making of the Atomic Bomb**, uses the misleading term "melt" to describe a fizzle.

> They could not use a gun to assemble a critical mass of such stuff: approaching each other even at 3,000 feet per second, the plutonium bullet and target would *melt* down and fizzle before the two parts had time to join. [emphasis added][83]

An effective bomb is expected to have at least a minimum impact, defined by the type of weapon. It was clear that the *Thin Man* would not likely achieve optimal performance, but the question remained as to whether it would even meet the minimum standard. At least one historic source from atomicarchive.com confirms the idea that the situation warranted an experiment.

> "The simplest way to proceed might have been to build a few experimental bombs in the early nineteen-forties and try them out. Not the least of the Laboratory's problems arose from the impossibility of doing this. By the time the precious shipments of fissionable material arrived at Los Alamos, a workable bomb design had to be ready. Various components and sub-assemblies could be tested by themselves, but no integral test of the weapon would be possible until long after the time when such testing might have served its purpose best."[84]

There have been two historic nuclear explosions that were considered fizzles. The first test at *Operation Crossroads*, Test *ABLE*, failed to meet expectations.

(The second, Test *BAKER*, far exceeded expectations; the radioactive fallout was unmanageable. In neither test were any ships "atomized", as was the case with the SS *E.A Bryan*, which was obliterated in the Port Chicago explosion. A third test, *Test CHARLIE*, was canceled.) The North Korea nuclear test of 2006 was also deemed a fizzle by many experts.

Disappointing. Fizzle. *Thin Man*? The prototype atomic bomb, the *Thin Man*, was shelved on July 17th.

In a subsequent letter to Bush, James. B. Conant, head of the National Defense Research Committee (NDRC), suggested the possibility of going back to the drawing board, if other options failed, to see if there was any way to increase the effectiveness of the Mark II. This seems to suggest that the Mark II was tested but did not meet expectations.

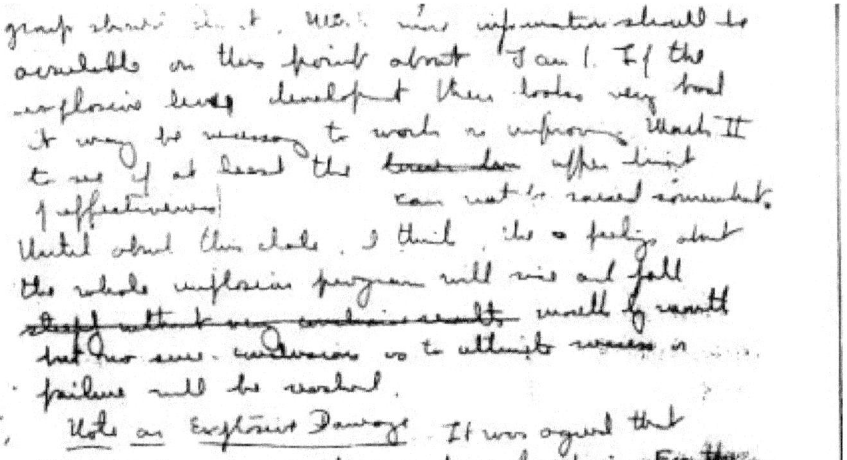
information should be available on this point about Jan 1. If the implosion being developed then looks very bad it may be necessary to work on improving Mark II to see if at least the upper limit of effectiveness can not be raised somewhat. Until about this date, I think, the feeling about the whole implosion program will rise and fall ... but then some conclusion as to ultimate success or failure will be reached.
6. Note on Explosive Damage. It was argued that Class B damage was damage beyond repair ...

Clip from Conant's original handwritten letter to Bush, August 1944.

Below, a redacted transcript.

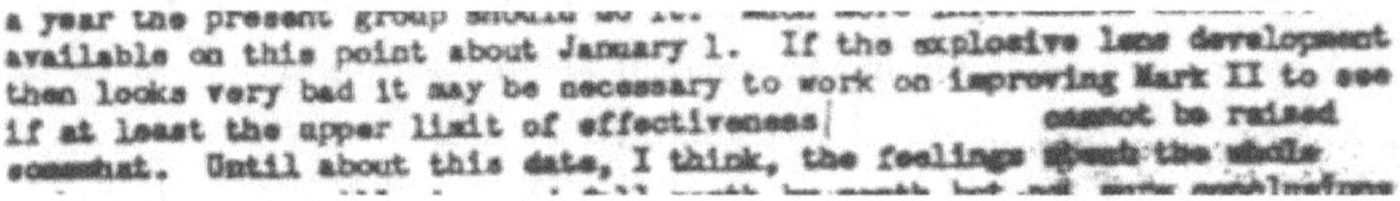
available on this point about January 1. If the explosive lens development
then looks very bad it may be necessary to work on improving Mark II to see
if at least the upper limit of effectiveness cannot be raised
somewhat. Until about this date, I think, the feelings

"... If the explosive lens development then looks very bad it may be necessary to work on improving Mark II to see if at least the upper limit of effectiveness [white out] cannot be raised somewhat..." [85]

In his book, **The Last Wave from Port Chicago**, Peter Vogel explored evidence that the device tested at Port Chicago was a deuterium hydride bomb designed for that occasion. My evidence suggests a closer inquiry into the fate of the *Thin Man*.

The disappearing Thin Man

The *Thin Man* bomb was 1) the first atomic bomb model ever produced; 2) the focus of the Manhattan Project for the first year; and 3) the prototype for the first atomic weapon used in warfare, the *Little Boy* uranium gun-assembly bomb.

Yet the *Thin Man* is the least-known of the four *named* atomic weapons created by the Los Alamos Laboratory during World War II (the *Thin Man*, the *Fat Man*, the *Gadget* and the *Little Boy*.) Unlike the other three famous bombs, there are no museum models of the *Thin Man*. Pictures show only the casings.

"Thin Man" plutonium gun test casings at Wendover Army Air Field, as part of Project Alberta in the Manhattan Project. "Fat Man" casings can be seen behind them. Wikipedia.

Through the lens of the PCnet, and in light of all the other supporting evidence, one reasonable explanation is that the *Thin Man* was tested at Port Chicago. This would explain both why there are no apparent remnants of the bomb and why it is so little-known.

The one thing that is emphasized in historic accounts is that the *Thin Man* would not work; could not work. The *implication* is that theoretical scientists reached this conclusion, and the ordnance team left it at that. According to science historian Lillian Hoddeson,

> The discovery by experimental physicists at Los Alamos in mid-1944 that the plutonium produced in a nuclear pile readily undergoes spontaneous fission . . . abruptly changed the course of history.[86]

Even Hoddeson found it noteworthy that (apparently) "... a purely scientific finding turned around the macroscopic program of one of the most consequential technological projects of our time."

Further study includes more than research, it requires analysis and interpretation as well. The premises of the standard narrative do not even raise the question, but the PCnet calls for a clear understanding of what is meant by "experimental physicists" and "purely scientific". After all, these terms do not necessarily rule out an ordnance experiment. In the first place, "purely" scientific research includes testing theories by experiments. As seen in the case of the shock wave, for example, the results of an underwater burst contradicted the predictions. (Which may have been disappointing.)

Even Einstein, in his work on torpedo design, reached a stage where he felt that nothing more could be done theoretically; experimentation was necessary, he said.[87]

Besides, the *Thin Man* was an experimental model, made to be tested. And Parsons was an Experimental Engineer who, if anything, overemphasized the need to conduct tests, early and often.

There seems to be no evidence of the *Thin Man*'s existence today—apparently no museum exhibit or other remnants. The actual bomb may have been created by the summer of 1944, which was the original date projected for its use.

The design for the prototype did not require a reusable weapon; rather, it called for one that could self-destruct in the explosion. *Thin Man* and *Fat Man* dummies were undergoing air drop tests. But the fate of the actual weapon is unclear. It was "shelved." What does that term mean exactly? Was the *research* put on hold, or was the *Thin Man* bomb literally put on a shelf? If so, where is it now? If not, what happened to it?

Even if they were certain the *Thin Man* would not "work" as desired, it would have had some kind of effect, which would have provided some kind of data, which is more than they would have had without it.

As noted previously, the predetonation problem was evident right from the beginning. It was discussed during the Los Alamos pre-opening conference in April 1943; but they made the decision to work on the plutonium gun first anyway because, for one thing, plutonium, a manufactured byproduct of uranium, was cheaper and available. If the more difficult design worked at all, it would provide the necessary proof of a uranium gun-assembly bomb, the *Little Boy*.

The *Little Boy* allegedly was used at Hiroshima without ever having been tested. Here, again, is an inexplicable "fact" in the traditional history. What was the basis of the confidence that the first atomic bomb would work?

The *Little Boy* was dropped on Japan a few weeks after the Trinity test. Based on the way the history is reported, it is easy to make the false assumption that the Trinity test – which is known as "the world's first nuclear explosion" – provided the proof that the *Little Boy* would work. But the Trinity test was the first *implosion* bomb. It was a plutonium bomb, a very different weapon from the *Little Boy*, in material as well as in design.

With all these facts in mind, we return to Parsons and his "observation" of the Port Chicago explosion.

In a September 1944 letter to Groves, Parsons used the phrase, "From my observation of the Port Chicago explosion . . ."

The seemingly naked phrase is stunning first because it seems to suggest that Parsons witnessed the blast, and secondly because historians seem to have overlooked the implication.

In any case, the phrase raises several questions for the PCnet. First, there is the question of semantics. If Parsons meant "from my *study of the effects* of the explosion," he used an odd choice of words. Otherwise, though, what other interpretation can there be? Could he have actually observed the explosion? If so, how? Why would he have been in a position to see it?

Christman found nothing in the Parsons' files to substantiate the possibility of a nuclear explosion at Port Chicago. He insisted that the theory was scandalous, that it had no merit whatsoever. At the same time, though, he cited Badash and Hewlett, who thought a nuclear test at Port Chicago would have been a heinous crime:

> Conducting a test in which the effects of conventional explosives would mask the effects of the nuclear weapon likewise shows little serious intent to study the phenomena. Choosing a major port that was in active use during intense fighting in the Pacific, rather than some isolated atoll (such as Bikini), is also inexplicable. Even the argument that the bomb was to be delivered by ship to a port, as mentioned in the famous letter Albert Einstein wrote to President Roosevelt in 1939, is historically inaccurate. Once tactical preparations began, the Army Air Corps, not the Navy, practiced delivery—and it would have been hard-pressed to carry the bomb before the spring of 1945 (Jones 1985, 519-20). In short, if the Port Chicago explosion was a deliberate test of a nuclear weapon to see the extent of damage on a port, not only was it poorly planned, it was a case of sabotage, treason, and murder.[88]

There is no way Christman's hero would have been guilty of such a crime. But Parsons himself would have likely objected to the historian's characterization, (just as he would have joined fellow officers in objecting to characterization of the attack on Hiroshima as a war crime, or even as immoral or unnecessary.) If he did conduct a test at Port Chicago, it would have been because he deemed it necessary in order to produce the weapon that would win the war.

Furthermore, the assumption that the Port Chicago explosion shows exactly how it would have masked a nuclear detonation, providing the perfect cover for a nuclear explosion. In fact, 364 days later, General Groves used a false report of an explosion at a remote ammunition magazine as a cover for the Trinity test.

As far as studying the phenomena, it might have been possible to perform calculations that would account for the high explosives on hand. In his preliminary memo on the Port Chicago data, Parsons reported a *limited* number of high explosives involved:

> "Amount of high explosive involved: 1500 to 1750 (2000 lb.) tons, of which about 200 tons were in cars on the dock. This is limited (by me) to TNT and torpex in bombs and depth charges, and does not take into account H.E. in projectiles and AP bombs, smokeless powder and incendiary material."[89]

Beyond physical damages, the atomic bomb was (and is) a psychological weapon. To Parsons, a successful test of a nuclear weapon would have to demonstrate its horror, including its impact on people. As noted earlier, there may have been a chilling explanation for Parsons' concern that foreign observers might find a "desert shot" disappointing. In his letter to Groves, he explained that

> "The principal difficulty with such a demonstration is that it would not be held one thousand feet over Times Square, where the human and material destruction would be obvious, but in an uninhabited desert, where there would be no humans and only sample structures."[90]

No amount of theoretical science could have provided that type of data.

Again, Badash and Hewlett's objections are readily overcome. The first one carries its own clue; a test using conventional explosives would have indeed *masked* the effects of the nuclear weapon. That is evident from the false press release Groves used to avert public attention away from the Trinity test. The claim was that there had been an explosion at a remote ammunition dump.

The test would have been set up in a way that took the conventional explosions into account. The base was designed to prevent *secondary* effects, results that would skew the data, as was the case with studies of the Halifax explosion.

The suggestion of the Bikini atoll suggests postwar sensibilities, and shows a lack of awareness about wartime needs and limitations. To conduct the secret test at an isolated atoll far from the continental U.S. during the war would have required unmanageable logistics, including setting up a complete base, staffing it and keeping it secret. On the other hand, the use of a major port that was in active use would demonstrate the real-world power of the weapon. During the secret research and development of the weapon, a desirable test location would have to be secure and remote. Port Chicago met both criteria.[91]

Parsons, who was in charge of Project Alberta, the division of Los Alamos that prepared for delivery of the bomb, had the cooperation of the Army Air Force. He was in charge of the tests where pilots were trained to test dummy bombs, called pumpkins. In March 1944, he arranged to move the tests from the Dahlgren Proving Ground in Virginia to MUROC (now Edwards Air Force base) in San Bernardino County, California, where the Fourth Air Force appears to have been assigned from March 31st, 1941 to *July 17th, 1944.*[92] There was a problem with the tests, so they stopped for a period in the spring, but resumed in June.

In a June 1944 memo, Parsons referred to "...My next trip west...". That may explain his absence from the meeting on July 17, 1944, when Manhattan Project leaders gathered at the University of Chicago and decided to shelve the *Thin Man* plutonium gun-assembly bomb. Parsons, head of ordnance, did not attend the meeting. His absence is not mentioned in the minutes.

The ordnance division had been working on the *Thin Man* for over a year, but they had not been able to overcome the problem of predetonation. They had discussed the problem in April 1943 at the conference held before the Lab opened. Plutonium had a quirk. Under certain circumstances, a natural chemical reaction could cause "spontaneous fission". In the same way spontaneous combustion can cause a fire to erupt, spontaneous fission in a gun assembly weapon could cause the plutonium to detonate too soon. This was not the case, though, with uranium.

Plutonium was manufactured, and therefore more readily accessible than uranium, which had to go through elaborate processing. If there was any way to make the gun work, great. But if not, at least the less expensive plutonium gun would provide the data needed for the uranium gun. Another problem was resolved when they realized it was not necessary to build a sturdy weapon, just one that would "hold together for one shot."

Traditional history says the *Little Boy* bomb went into combat without ever being tested because the scientists were confident that it would work. But Parsons was the head of ordnance; and he was a stickler for testing. He believed nothing worked until it had been done at least once. The *Thin Man* gun-assembly weapon was the first atomic bomb model, predecessor of the *Little Boy* uranium bomb and proof that it would work.

So, the question is, did the *Thin Man* prove the *Little Boy* would work? If so, how?

From a letter Parsons wrote to Groves in June, along with a series of memos he wrote during that period, it appears that he was preparing for a forthcoming experiment on the West Coast.

SECRET

UNCLASSIFIED

THIS DOCUMENT CONTAINS 1 PAGES
THIS IS COPY 2 OF 3

310145

353.4

P. O. Box 1663
Santa Fe, New Mexico
June 1, 1944.

Major General L. R. Groves
P. O. Box 2610
Washington, D. C.

R

Dear General Groves:

As you know, out setup for conducting experimental tests on the West Coast is under the Fourth Air Force, commanded by General Lind, with headquarters in San Francisco. I understand that General Lind has made frequent changes, and perhaps will continue to make these changes, in the commanding officers of the various air bases and depots on the west Coast. It would appear from events of the past several months, that the continuity in that command is at the top rather than in any given air base. Also, the directive signed by General Echols was sent to the Commanding General, Fourth Air Force.

In view of the above, and since we require cooperation and rapid action, I believe that it is desirable that I call on General Lind and invite him to witness at least one of the tests at Site W. If you agree that this action is desirable, I suggest that you arrange clearance for it, so that I may make the call during my next trip West, which I now expect will begin 12 June.

Sincerely,

W. S. Parsons
Captain, USN

WSP/hg

cc: Mr. Oppenheimer

FINAL DETERMINATION
UNCLASSIFIED
L. M. Redman
OCT 28, 1950

"As you know, our setup for conducting ***experimental tests on the West Coast*** *is under* ***the Fourth Air Force****, Commanded by General Lind, with* ***headquarters in San Francisco*** *. . . In view of the above, and since* ***we require cooperation and rapid action****, I believe that it is desirable that I call on General Lind and* ***invite him to witness at least one of the tests*** *at Site W. If you agree that this action is desirable, I suggest that* ***you arrange clearance for it****, so that I may make the call during* ***my next trip West, which I now expect will begin 12 June****."*

Parsons to Groves, June 1, 1944 *[emphasis added]*[93]

A straightforward reading of this document, with no attempt to detect any coded language Parsons may have used, says he planned to use the Army Air Force in forthcoming experiments on the West Coast, where he would be from around June 12th until an unspecified date. General Lind should be invited to witness at least one of the tests, which presumes at least one of them would be noteworthy, for whatever reason, and of interest to the general.

Viewed through the lens of the PCnet, this memo raises several questions:

- Did Parsons use the Fourth Air Force routinely, or was this a special assignment?
- Why did he use the general term "experiments on the West Coast" instead of stating a specific location?
- Was there something special about the tests that would warrant General Lind's interest?
- Why did the General need clearance to witness a test involving his own unit, the Fourth Air Force?
- What type of experiment called for "rapid action" on the part of the Fourth Air Force?

These questions should be answered in view of other communications from Parsons around the same time.

July 4, 1944.

Messrs. Bonbrake, Ramsey, Mitchell

W. S. Parsons

CLASSIFICATION CANCELLED
PER DOC REVIEW JAN. 1973

FAT MAN DUMMY FOR TESTS AT Y; SHIPMENT OF

On 3 July I discussed shipment of the fat man dummy with type A tail, which is being held in Detroit for shipment to Y. I instructed Lt. Moore to make arrangements to ship this dummy and tail, separately boxed, by truck to Mr. C. H. Schruben, 210 East Ohio Street, Chicago, Illinois.

It is requested that Mitchell's office inform Mr. Schruben in regard to this shipment, which he should receive within the week and which he should then re-ship by the next freight car which leaves Chicago for Y.

It would be well to inform Mr. Schruben that the boxes in this case are approximately six foot cubes and that the heavy one weighs about nine thousand pounds.

WSP/hg W. S. Parsons

Parsons to Bonbrake, Ramsey and Mitchell, July 4, 1944. In July 1944, Parsons wrote at least two memos regarding the shipment of one particular "Fat Man dummy". Besides use of that name, it is unclear how much of the memo is in code.

These messages contain at least a bit of secret coding, a common Manhattan Project practice. For example, at the time of the Trinity test, July 16, 1945, President Truman was in Germany at the Potsdam Conference. Secretary of War Henry Stimson used medical language to inform President Truman that the test had been successful: "Operated this morning. Diagnosis not complete but results seem satisfactory and already exceed expectations . . . Dr. Groves pleased."

"Fat Man Dummy" was a code name for one of the model bombs used during tests that served the double purpose of training pilots and proofing the hardware of the various mechanisms involved in the new technology. But why was Parsons personally involved with the delivery of this one particular "Fat Man dummy"? What was so special about this one delivery that it required special handling?

UNCLASS

July 14, 1944.

…ers. Ramsey, Bainbridge, Mitchell

W. S. Parsons

<u>Fat Man Dummy; Delivery from Chicago</u>

While in Chicago I checked on the whereabouts of the fat man dummy which is on its way here from Detroit. Mr. Schruben told Captain Jones that it would be shipped as machinery in the next carload lot coming to the engineer warehouse from Chicago. Mr. Schruben predicted that this shipment would arrive Santa Fe 17 or 18 July.

Since the heavy box weighs about 9,000# and is a cube, with about 62" on a side, the unloading from the freight car and loading on a truck will require special handling facilities.

It is requested that this matter be gone into, and preparations made before attempting to unload the fat man.

WSP/hg

W. S. Parsons

In this July 14, 1944 memo about the "Fat Man dummy",

Parsons uses the phrase "while in Chicago" and "on its way here".

An earlier memo places him on the West Coast starting in mid-June.

One *<u>possibility</u>* is that Parsons could have used "Chicago" as a code for "Port Chicago". Since his next trip to the West Coast was to begin in the middle of

June, Parsons may have still been there when he wrote these memos in July. The date of the shipment, "17 or 18 of July", could refer to the date scheduled for the detonation.

In a different context, this type of speculation might be a bit far-fetched, or contrived; but this was the very style of communication that went on in the Project. If there was a secret test at Port Chicago, written communications would have been avoided; if absolutely necessary, they would have certainly been coded.

The PCnet facts considers these facts in light of various statements by Parsons:

- "Port Chicago was designed for large explosions."

- "From my observation of the Port Chicago explosion . . ."

- ". . . until we have done something once, we cannot be sure it will work."

Parsons was determined to deliver the perfect weapon. After the *Thin Man* was shelved—on the day of the Port Chicago explosion— the implosion bomb became the focus of research and development at the Lab.

The Port Chicago explosion and the Manhattan Project

Connections matter.

There is at least one simple, obvious and non-controversial explanation for the overlapping timeline of Port Chicago and the Manhattan Project: they may indeed have been two separate responses to wartime necessities.

On the other hand, the unexplored but documented connection between the explosion and the creation of the atomic bomb warrants further investigation. It is beyond the scope of this book to probe deeper into the logical *possibility* that the PCNAD was designed to serve as the location for a test of the weapon under production at the Los Alamos Laboratory.[94] Still, it is evident that the war provided the singular opportunity to build and use a weapon that would, and did, change the world.

In any case, there is more to these overlapping facts than what shows up in the traditional history of the bomb. For example, historians have only recently begun to examine FDR's decision to create the weapon, to assess his motives and plans.

It seems also significant that before becoming head of ordnance for the Manhattan Project, Parsons had served as Special Assistant to Vannevar Bush, who recommended him for the job. Parsons was a powerful man with widespread connections who enjoyed unusual authority and autonomy. The fact that he was close to Bush, who had the President's ear and was the man behind the Manhattan Project, further underscores this fact.

Again, Parsons made several important contributions to the war and helped shape post-war nuclear history. Yet he remains hidden on the unexamined timeline shared by the Manhattan Project and the Port Chicago explosion.

Parsons' obscurity is due in part to the myth that the Navy was excluded from the Project. Parsons was not the only person from the Navy to serve on the Project; he himself brought many aboard. If there was a secret nuclear test at Port Chicago, one way to divert attention from it would be to divert attention from the contributions of the Navy.

Several facts about the Port Chicago Naval Ammunition Depot itself also support the PCnet. For one thing, the base was established at the time the Project was getting underway. Data from the explosion at Port Chicago was critical to the creation of the bomb, and it was not obtainable by any other means. Up till now, Peter Vogel was the only person to examine this critical link at any length. The updated version of the PCnet, as presented in this book (and other work by this author) addresses this historical oversight, encouraging and inviting the scrutiny of other researchers.

If there was a need to create an atomic bomb, and a corresponding need to test the unprecedented weapon of unconceivable power, then there was a need for a remote location where the physically dangerous, politically hazardous and economically risky test could be conducted in secret. There was no better time than during the war to create the bomb, no better place to test it than the Port Chicago Naval Ammunition Depot.

The PCNAD would have met the criteria for a remote, controlled location to conduct a secret test.

Again, assuming that it did take place, the experimental explosion at Port Chicago was protected by the secrecy of the Manhattan Project, by the secretive nature of the President and his men, and by the silence of the Press. Additionally, the Port Chicago Naval Ammunition Depot (PCNAD) itself had a built-in, multi-layered cover:

- Although the Port Chicago disaster involved one of the largest explosions in history, the story of the "unexplained" explosion was soon and permanently trumped by the story of the controversial Port Chicago mutiny.

- The Navy was supposedly left out of the Project, but the head of ordnance was a Navy captain and many naval officers played key roles in the Project. (Otherwise, history requires an explanation for why a Navy captain served as weaponeer aboard the *Enola Gay* and was one of the three "Chiefs of Staff" at Tinian, where the bombing of Japan was staged.)

• The PCNAD was a remote military location that was off limits to the public and gave only limited access to the press.

• The men planning the research and development of the bomb realized from the very beginning that when it came time to test the bomb, a remote facility would be needed. The new remote ammunition magazine at Port Chicago was established in 1942, the same year the Manhattan Project became official, and the same year the Navy opened all ratings to African American enlistees.

• All military personnel are "expendable" during a war. Given the avoidable hazardous conditions that went unaddressed at Port Chicago, it is evident that there was little concern for the safety of the allegedly "low quality personnel" who staffed the ammunition magazine. Consequently, the PCnet takes particular note of the coincidental timing of the Navy opening its ranks to Negroes. Camp Robert Smalls, the segregated camp at the Great Lakes Naval Training Center near Chicago, Illinois was established just in time for Negro sailors to be available to serve at the Port Chicago when the base opened in December 1942. As discussed earlier, this also coincides with FDR's change of heart about Negroes in the Navy, and his ideas about where they could serve, without integrating ships

A "confidential board" called for the establishment of the PCNAD.

Port Chicago, of course, does not show up on any official timeline of the Manhattan Project. The overlapping timelines of the Naval magazine and the bomb project become both noticeable and significant only when the history is viewed from the perspective of the PCnet.

The official history of Port Chicago says it was constructed pursuant to the December 1941 recommendation of "a confidential board." But who recommended what, and why?

Even decades after the war, the specific person or agency making the recommendation is not identified in historical accounts. This lack of accountability is so unusual as to be glaring. In typical reports of this nature, enough information is given for a researcher to determine precisely who held ultimate responsibility for the decision.

(Examples abound. In other historical accounts of naval establishments, the agencies are identified in the reports: *the Navy Department; the Bureau of Yards and Docks; the Hepburn Board; the Fuel Storage Board; the Chief of the Bureau of Navigation; the Commander in Chief of the Atlantic Fleet . . .* to name a few. In this rare case, the recommendation came from an unspecified, confidential "board of officers of the Twelfth Naval District.")

Although the vague wording does not assign clear responsibility to him, the statement *suggests* that the board convened and made its recommendation under the authority of the Commandant of the Twelfth Naval District, Rear Admiral Carleton E. Wright. Admiral Wright otherwise figures into the Port Chicago story in several ways.

Perhaps most importantly, Wright issued the *July 14th, 1944* precept for the Port Chicago court martial. Immediately after the explosion on *July 17th*, he praised the Negro sailors who helped put out the few fires caused by the blast; they performed like true sailors, "as was to be expected." Weeks later, Wright addressed the sailors who refused to go back to work. Rebuking them as cowards, he said he would see to it that they got the death penalty if they did not go back to work.

On July 20, 1944, Wright engaged in a brief conversation with Parsons at the Mare Island Naval Shipyard as he headed to Port Chicago, accompanied by his brother-in-law, Cpt. Jack Crenshaw, one of the three men on the Court of Inquiry that investigated the explosion.

It would require further research to determine who was on the board that recommended the establishment of Port Chicago. What prompted the board to convene? What were the factors they considered, and what were their points of disagreement? This missing information is standard in such historic reports. This oversight is one of many brow-lifting omissions in the history of the bomb. This omission is of special interest because of 1) the interconnections between wartime policymakers; 2) Parsons' autonomy and his link to men in high places; and 3) the overlapping timelines of the Project and PCNAD.

Certain coincidences raise questions:

FDR approved EO 8807, establishing the Office of Scientific Research and Development (OSRD) on June 28, 1941. The OSRD had oversight of the Manhattan Project. Bush moved up from the National Defense Research Committee (NDRC) to head the new agency, which had broader authority. His close colleague, James B. Conant of Harvard University took over leadership of the NDRC. Three days before approving the OSRD, FDR signed EO 8802, establishing the Fair Employment Practices Committee, to avert the threat of a Negro March on Washington.

On January 19, 1942, FDR gave Bush the official go-ahead to proceed on the bomb project. That April, the Navy began accepting Negro recruits. They were trained for general ratings, but most were assigned to shore duty, including at Port Chicago, where all the ship loaders were Black.

Establishment of the Port Chicago Naval Ammunition Depot at the same time as the OSRD, the agency that administered the Manhattan Project was, by definition, coincidental. But was it *just* a coincidence?[95] Only further research can establish the answer to that important question, which is raised only by investigation of the PCnet.

Designed for large explosions

Information about the board that recommended the establishment of the PCNAD is still confidential.

According to one official history, there were disagreements about the *purposes and design* of the base.[96] The report does not identify the people involved, nor does it reveal either their separate objectives or the nature of their disagreements. In general, though, naval magazines were set up to produce and store ammunition. Port Chicago, established in a remote location on the San Francisco Bay, also shipped ammunition to the warzone.

Further research might disclose who was on the confidential board. It is less likely, though, that *typical* historical research, in archives or even memoirs, will upturn any documents that reveal what their priorities were.

After all, what was there to debate about? The type of structures to be built? The layout? Staffing? The need to fortify the base? Its location?

Surely there was no disagreement about the rationale for building the base. The official story is that Port Chicago was created to provide support for nearby Mare Island, shipment of arms to the South Pacific.

The U.S. had not yet entered the war, but military preparedness was one of FDR's steadfast goals. As Assistant Secretary of the Navy during World War II, he had pushed for a stronger Navy. As World War II approached, the Navy addressed the need for various new facilities. As a consequence, "Expansion of the Navy's ammunition depot facilities had been under way for a year and a half when war became an actuality."

Port Chicago was "an accident waiting to happen." A nuclear test at Port Chicago would have been readily camouflaged by 1) the large amount of conventional ammunition on the pier, which included high explosives; and 2) the known hazards at the base, which, according to testimony, including untrained and inexperienced officers racing untrained (and low quality) personnel. Though many ranking officers were aware of the hazardous conditions, they went unaddressed. Why?

The **Bluejacket Manual** distributed to Navy enlistees during the war says, "There are no accidents." People made accidents; accidents were avoidable. So, if the brand-new facility was so crucial to the war, why wasn't safety a higher priority? The PCnet explores the idea that the appearance and threat of an accident may have been cultivated on purpose as a cover for the nuclear test and to distract attention from the explosion itself.

Historians "interrogate" and interpret historical documents.

Historians, of course, have not addressed the possibility that the PCNAD was "an accident waiting to happen" *by design*. After all, they have not considered the history from the PCnet point of view, only from that of the standard (flawed) history. In **The Color of War**, for example, James Campbell *interprets* Parsons to mean that Port Chicago was designed to minimize damages *in case of* a large explosion.

Accurate history requires correct interpretation of primary resources. Historians "interrogate" historical documents. They ask questions, for example, about the author, dates, any markings, the chain of command, and so forth. To understand the meaning of a document, its relevance and significance, it is also necessary to study the wording, and to do so in the context of the author's language, experience and intent.

There are two problems with Campbell's interpretation.

1. First of all, Parsons' straightforward phrasing was "... Port Chicago was designed for large explosions." Campbell translates "*for*" as "*in order to avoid*". Through the lens of the PCnet, this interpretation is the inverse of what Parsons said. As a Naval officer, a liaison both to Vannevar Bush and to the Navy Bureau of Ordnance, Parsons knew how to choose his words. As an exemplary Experimental Engineer, he was interested in, worked at and helped design facilities that were designed to withstand experimental explosions.

1. Secondly, to "minimize damages" only *implies* concern for safety; it does not necessarily mean accident prevention. After all, the best way to minimize *all* damage from an explosion is to prevent one in the first place, by establishing and enforcing preventative safety measures.

As indicated earlier, Parsons noted that "with luck" the fortified design of the PCNAD may have prevented 'secondary' fires. Since the data from the explosion was used to estimate expectations from a nuclear detonation, it would not be mere speculation to conclude that those damages were what he considered 'primary'. However, historical accuracy demands that instead of making that reasonable assumption, the question should be researched. Who knows what additional facts might surface in the investigation?

One way to *assure* an accident is to neglect known hazards. What if Port Chicago was actually managed according to design? As Parsons indicated, his interest in visiting the scene was to study the effects of the "detonation". The design of other proving grounds provided for the detonation of a powerful weapon.

Above all, Parsons' statement must be viewed in the context of his role and his mission. For his purposes, any *secondary* damages would skew the data.

Still and all, Campbell's interpretation, even seen from a traditional point of view, takes nothing away from the idea that Parsons was speaking literally. "Port Chicago was designed for large explosions."

An accident designed to happen

One way to conceal a secret explosion would be to situate it at a site where an accident would be expected. Although the Court of Inquiry said the cause could not be determined, the narrative that developed attributes it to an accident caused by the black sailors. That erroneous assumption is logical; however, it is also convenient.

Sources recite well-established 'facts' – cherry-picking and sometimes even distorting them to support the accepted narrative. Less obvious facts are simply ignored; the questions they would raise remain unasked. Instead of jumping to conclusions, the PCnet requires close examination of the evidence.

Christman insisted that the Port Chicago explosion was not a major factor in the research and development of the bomb. But according to Parsons, the data from the Port Chicago explosion marked a turning point in the Manhattan Project. In a post-war speech to the Naval War College, he explained that:

> "The Naval magazine at Port Chicago had built-in protection against large explosions. *With some luck, this protection was effective in preventing any secondary explosions or fires. In the light of the Port Chicago results, we reanalyzed the Halifax result."*[97] [emphasis added]

The best data available was from the Halifax explosion; but it was unusable.

According to Parsons, the Halifax explosion was discussed by the core Manhattan Project personnel who attended the April 1943 conference before the Los Alamos Lab opened.

In fact, Halifax had been discussed two years earlier, when the NDRC was still exploring the feasibility of the bomb. The National Academy of Science (NAS) studied the problem. In a November 1941 letter to Bush, chairman

Frank Jewett explained that the Halifax explosion would make a poor comparison for atomic bomb research. According to Jewett,

> . . . the case occasionally cited in [sic] the Halifax explosion of a munitions ship during the last war. The damage there extended to a radius of much more than a mile but in the opinion of the writer this is *not a fair case to consider because the conditions for larger damage were unusually favorable* in that the exploded ship was in the middle of the harbor and the city forms a sort of a natural amphitheater with no barriers to reduce the effects of the blast. [98]

the former, the case occasionally cited in the Halifax explosion of a munitions ship during the last war. The damage there extended to a radius of much more than a mile but in the opinion of the writer this is not a fair case to consider because the conditions for large damage were unusually favorable in that the exploded ship was in the middle of the harbor and the city forms a sort of a natural amphitheatre with no barriers to reduce the effects of the blast. The land mines

In this November 1941 report, the National Academy of Scientists explained why data from the deadly Halifax explosion would not be useful in the design and production of the atomic bomb. This is why the Port Chicago explosion data was critical to the Project. [underline added]

Right from the beginning, Manhattan Project leaders knew they had a future problem to address. When it came time to evaluate the effects of the bomb, they would need a fortified, secure, remote location, one they could control. And they could not wait for it to materialize; they had to get ready to deal with it.

There would be damages from the test of an atomic weapon; in fact, that would be the very point of a developmental experiment. Parsons' job was to design and conduct tests of new military technology. He pushed for the development of new military technologies, and was willing to take on risky, unproven ideas.

As an Experimental Engineer, Parsons understood why no new technology was reliable until it had been shaken down. That philosophy is demonstrated by his hands-on work on the proximity fuze. Christman describes the April 13, 1942 experiment where a gun crew fired projectiles with proximity fuzes against airplanes. Their twenty percent success rate convinced Parsons that

> " . . . . the time had come for the big test: firings from standard navy guns by regular gun crews from a real ship against attacking aircraft. No ordinary navy commander would have had the power to commandeer a new light cruiser fresh from sea trials for an ordnance test. But by now Parsons was no ordinary field-grade officer. Because of a unique combination of personality, reputation (military and scientific), and circumstance, he had the backing of the biggest guns of the uniformed navy—Admirals Ernest King and Spike Blandy—as well as the president's science czar, Vannevar Bush. Thus, the crew of the USS *Cleveland,* about one thousand all told, found themselves in the summer of 1942 off of Norfolk, Virginia, preparing their 5-inch antiaircraft batteries for secret tests."[99]

Parsons also obtained four drones to be used as targets. (He wanted six but had to settle for four.)

That fall, Parsons insisted on taking the new technology into combat himself. In preparation, he arranged for five hundred fuzes to be stored at Mare Island. In addition to his resourcefulness and his access to unlimited resources, the elaborate system Parsons established shows how he would have ensured the success of an experiment at Port Chicago. Significantly, Christman adds that

> "Four hundred and fifty of these shells were then set aside in a special stockpile under his name. The other fifty shells, complete with fuzes, were flown back to Dahlgren, where they were tested to assure Parsons that no new problems had crept into the process. **He followed the same procedure on all the succeeding shipments.**"[100] [emphasis added]

Parsons would likely have followed a similar procedure at Port Chicago. That would explain why his name appears on some boxes included in the Daily Inventory of one of the freight cars still intact after the July 17, 1944 explosion.

The proximity fuze was a new weapon; the atomic bomb was a new *type* of weapon. Everything about it was unknown. It would be hard enough to translate data from a conventional explosion into useful information on how to design a uranium bomb. The greater challenge would be to figure out how much uranium was needed and to predict the magnitude and effects of the unprecedented weapon.

All this, and more, figured into research on how to configure, detonate and deliver the bomb. There simply was no recipe that could tell them how to bake the "yellow cake", (impure uranium oxide obtained during processing of uranium ore.) But bake it they must; they had to find a way.

The bomb had to be built, and they had to find a way to do it. It had to be studied, designed, created and its power had to be demonstrated before the end of the war. There was no time for shortcuts.

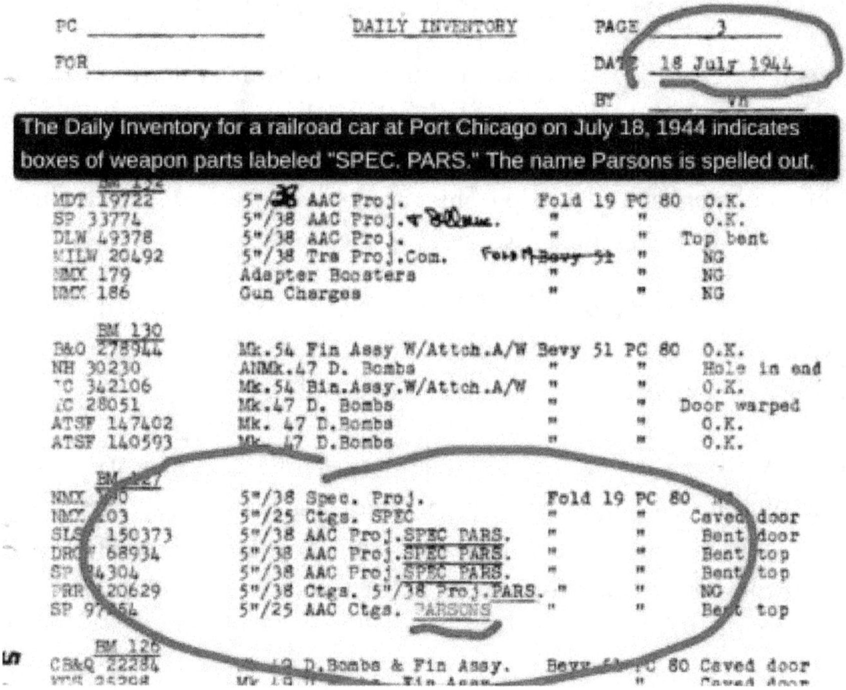

PC ______ DAILY INVENTORY PAGE 3

FOR ______ DATE 18 July 1944

BY [illegible]

The Daily Inventory for a railroad car at Port Chicago on July 18, 1944 indicates boxes of weapon parts labeled "SPEC. PARS." The name Parsons is spelled out.

MDT 19722	5"/38 AAC Proj.	Fold 19	PC 80	O.K.
SP 33774	5"/38 AAC Proj.	"	"	O.K.
DLW 49378	5"/38 AAC Proj.	"	"	Top bent
MILW 20492	5"/38 Trs Proj.Com.	~~Bevy 51~~	"	NG
NMX 179	Adapter Boosters	"	"	NG
NMX 186	Gun Charges	"	"	NG
BM 130				
B&O 278944	Mk.54 Fin Assy W/Attch.A/W	Bevy 51	PC 80	O.K.
NH 30230	ANMk.47 D. Bombs	"	"	Hole in end
[illegible] 342106	Mk.54 Bin.Assy.W/Attch.A/W	"	"	O.K.
[illegible] 28051	Mk.47 D. Bombs	"	"	Door warped
ATSF 147402	Mk. 47 D.Bombs	"	"	O.K.
ATSF 140593	Mk. 47 D.Bombs	"	"	O.K.
BM [illegible]				
NMX [illegible]	5"/38 Spec. Proj.	Fold 19	PC 80	[illegible]
NMX [illegible]03	5"/25 Ctgs. SPEC	"	"	Caved door
SLS[illegible] 150373	5"/38 AAC Proj.SPEC PARS.	"	"	Bent door
DR[illegible] 68934	5"/38 AAC Proj.SPEC PARS.	"	"	Bent top
SP [illegible]4304	5"/38 AAC Proj.SPEC PARS.	"	"	Bent top
PRR [illegible]20629	5"/38 Ctgs. 5"/38 Proj.PARS.	"	"	NG
SP 97[illegible]4	5"/25 AAC Ctgs. PARSONS	"	"	Bent top
BM 126				
CB&Q 22284	[illegible] D.Bombs & Fin Assy.	Bevy [illegible]	PC 80	Caved door
[illegible]	[illegible] Fin Assy.	"	"	Caved door

The July 18, 1944 inventory of a train at Port Chicago lists several boxes labeled with ordnance parts and the name "Parsons". (Document courtesy of Peter Vogel.)

Deak Parsons, cowpuncher

William S. "Deak" Parsons was an Experimental Officer both by nature and by calling. As a child, he conducted independent experiments to see what would happen to strips of metal after he laid them on the railroad tracks near his home; he enlisted the help of his sister, Clara. As a tall, slender, balding officer in his early 50s, he was still very goal oriented.

Besides his declaration that Port Chicago was designed for large experiments, Parsons made several statements that, each and all, warrant investigation of the PCnet. Here is a brief sampling:

- With regard to the necessary testing of the proximity fuze, which he took into combat for their first real-war test, he said **"There are enough new features to our problems to justify that until we have done something at least once we cannot count on it being done."**

This statement by the man in charge of developing and delivering the bomb contrasts with the common wisdom that says the *Little Boy* bomb – the first nuclear weapon ever used in a war setting – was never tested.

- In September 1944, Parsons wrote a letter to J. Robert Oppenheimer, Los Alamos director, urging him to do whatever was necessary to get the job done. "Ruthless men" must not let anything get in the way of developing "our bomb".

The statement made it into the lyrics of the 2005 opera, "*Dr. Atomic*"— where, of course, it is *not* attributed to Parsons. As in the case of nuclear history and contemporary nuclear policy, Parsons' impact prevails, but his cover remains intact.

Ruthlessness was a trait occasionally championed by Vannevar Bush and President Roosevelt, as well by Parsons. Other statements in the same letter reveal more of his beliefs:

• "Some tender souls are appalled at the idea of the horrible destruction which the bomb might wreak in battle delivery. I believe these imaginations magnify the horror beyond what they read about in thousand-bomber raids."

• Using the words "Based on my observation of the Port Chicago explosion," Parsons insisted that foreign observers would be disappointed with "a desert shot." He felt that "regardless of the scientific achievement it would be "a political and military fizzle."

• "The principal difficulty with such a demonstration is that it would not be held one thousand feet over Times Square, where the human and material destruction would be obvious, but in an uninhabited desert, where there would be no humans and only sample structures."

• Parsons was concerned that production of new military technology would be stymied after the war. In peacetime, bureaucratic procedures would again "strangle new ideas.

• "The *unity* that made the proximity fuze and the atomic bomb possible would no longer exist. And in the absence of war there would be a temptation to believe our own propaganda and conclude that since we have the best anyway why worry about improving it by expensive research and development."[emphasis added]

This statement underscores his postwar admission (or boast) to the Naval War College that the U.S. had been under an efficient dictatorship during the war.[101]

• Describing the success of the attack on Hiroshima in a letter to a friend, Parsons wrote, "Once in many centuries you can't shake off the Midas touch.[102] That's what happened to us." Other members of the crew, who had not known they were delivering an atomic weapon, cried out, "Oh, God!" – as in, "what have we done?"[103] To Parsons, the mushroom cloud billowing over the city was "awesome."

- "At that moment I knew what the Japs were in for, but I felt no particular emotion about it." Most important to Parsons was the "awesome responsibility" of delivering the world's first nuclear weapon.

- As Deputy for Technical Direction of *Operation Crossroads*, Parsons overruled the warning by Dr. Warren Stafford that the hazards would be too great. A third test had to be canceled after the "*BAKER*" test spread radioactivity across the region.

Parsons was the first person to recommend a nuclear proving ground in the continental U.S. After 67 detonations, the Marshall Islands tests[104] were becoming too expensive. Although his recommendation did not go over well at first, Parsons was persistent. Ultimately following his advice, the U.S. established the Nevada Proving Grounds, the site where the Atomic Vets were deliberately exposed to nuclear explosions. Decades later, the Government acknowledged the hazards of the test and approved compensation for those veterans who could demonstrate that their health issues were a direct result of that exposure to radioactivity.

Other U.S. citizens have been harmed by nuclear tests and waste disposal, including the "Downwinders" of Utah (who were not eligible for compensation under the Radioactivity Exposure Compensation Act (RECA) of 1990, which was renewed in July 2022; and the citizens of St. Louis, Missouri, who were unaware of the plutonium carelessly stored in their city.

Parsons expressed annoyance with people's unreasonable fear of the bomb. He believed Americans should be okay with a nuclear site within 150 miles of their homes.

Parsons was on the advisory committee that made the initial recommendations on nuclear policy to the Joint Chiefs. Among their recommendations: 1) the U.S. must continue producing more and better nuclear weapons; 2) the Congress should redefine 'aggression' so as to empower the President to authorize a first strike against any nation that might have been considering an attack on the U.S. 3) The U.S. Government should be able to conduct nuclear tests anytime, anywhere.

In short, though Parsons is best known as "the Navy captain who armed the *Little Boy* bomb aboard the *Enola Gay*," his greater claim to fame should be the fact that he had the awesome responsibility of playing midwife to the bomb. He fulfilled his promise to General Groves: the perfect weapon with the perfect delivery.

One way to assess the likelihood of a nuclear explosion at Port Chicago is to consider Parsons' personality, power and passion: his ruthlessness, his passion for perfection, his keen interest in testing, and his insistence that nothing must stop the perfect delivery of the perfect weapon.

Parsons was a hands-on leader. He took time to visit Port Chicago in person, to study the effects of the "detonation." His preliminary report on the disaster is listed among the primary documents on the Atomic Heritage site.

On the surface, the document appears to be an internal memo from a Naval officer to his commanding officer. The ruse was possible thanks to his Navy cover, made possible by keeping him on the staff or Admiral Ernest King while he worked for the Manhattan Project – one of many forms of concealment that remain intact decades after the war.

If there was a secret test at Port Chicago, it would have been possible not only to cover it up, but to continue shielding the secret indefinitely. But for Vogel's accidental discovery of the historical document from Los Alamos, none would be the wiser.

To see what he could see

Historians offer too many explanations, based on little or no documentary support, for why Parsons went to the scene of the Port Chicago explosion.

Christman, Parsons' sole biographer, gives several explanations. According to him, data-gathering was the least of Parsons' interests. Rather, Chrisman suggests several benign motivations. Parsons went to the scene on his own, out of his own curiosity, and, as the Senior Naval officer at Los Alamos, out of concern for the sailors; and, yes, perhaps to see if the data might be useful in the research and development of the bomb.

Roger Meade, archivist at Los Alamos National Library and co-author of *Critical Assembly*, one of the acclaimed histories of the bomb, gave a lecture on Parsons as part of the LANL 70th Anniversary Public Lecture Series. Whereas Christman said Parsons didn't have to go to Port Chicago; Meade said he was sent. Neither historian provides sources for their assumptions.[105]

Speaking of the congenial way Parsons dealt with an inconsequential memo from Groves, Meade used the unfortunate term, "in typical Parsons' fashion". It seems he coined the phrase from the last line of the chart, "History of 10,000 Ton Gadget", the historic document that triggered Vogel's thirty-five-year quest to understand the link between Port Chicago and the Manhattan Project. It may be a standard phrase at Los Alamos; it is less likely that Meade's unfortunate use of the phrase was just an unrelated coincidence.

Although Meade and Christman both downplayed the importance of the explosion to the Manhattan Project, they also rationalized why the head of a team that was planning for a large explosion would want to see it for himself. But, as noted earlier, Parsons himself – mentioning the explosion more than three times in two paragraphs – stated that the data from Port Chicago was of great importance to the development of the bomb.

> "The Naval magazine at Port Chicago had built-in protection against large explosions. With some luck, this protection was effective in preventing any secondary explosions or fires. In the light of the Port Chicago results, we reanalyzed the Halifax reports and concluded that a major factor in the Halifax disaster was a record blizzard which struck the city a few hours after the explosion.
>
> The results of Halifax and Port Chicago have been confirmed on a larger scale by Hiroshima, Nagasaki and Bikini "ABLE."[106]

While faced with the urgency to create the first atomic weapon before the end of the war, the head of ordnance for the Manhattan Project took the time to visit the scene of the Port Chicago explosion in person. He himself wrote at least one "preliminary" memorandum summarizing what he learned from that visit. His team analyzed the results of the explosion, comparing the data to that of other nuclear detonations, including the attack on Japan. Following the same protocol used with scientific experiments, the data from one test was used to set up and evaluate the next one.

The significance of this fact is underscored by the experimental nature of the two atomic bombs used in the attack. The use of the two bombs was experimental with regard to their effects and capabilities as new military technologies but also as tools of diplomacy.

As Arjun Makhijani explained in "*Nuclear targeting: The first 60 years*", the bomb had to be used during the war in order to demonstrate its power and the corresponding status of the United States. "The first experiments in "proper use" were the bombings of Hiroshima and Nagasaki."[107]

Research

As right-hand man to Bush in 1941, Parsons was in position to have inside knowledge about the origin of the Port Chicago Naval Ammunition Depot. It is even possible likely that as an expert in explosives testing, he would have been consulted with regard to establishing a location for the tests that would be required for the research and development of the bomb.

Before he joined the Project, Parsons led the development and testing of the proximity fuze (proving himself as well as the vital new military technology.) In 1946, he was the technical director over *Operation Crossroads*, the first post-war nuclear tests. His recommendation for a continental site led to the Nevada Proving Ground.

No orthodox research is apt to reveal evidence of off-the-record conversations; but then, the *flawed* history of the bomb is due in part to reliance on orthodox research methods. In **The Age of Hiroshima**, Sean Malloy writes about the inadequacy of traditional research methods when talking about race and the bomb. His statement applies to other aspects of the historical narrative as well.

> ". . . all these analyses, including Takaki's, rely on a way of thinking about race and racism that is extraordinarily narrow and ahistorical. **That narrowness is in part a result of the way in which most scholars have approached the evidentiary record on this question. Diplomatic and military historians have traditionally been rooted in archival research and government documents**, and there is, at least on the face of it, little in the official record that gives scholars much traction on the issue of race and the bomb. [emphasis added][108]

The primary sources in traditional archives and agencies are limited; so is the perspective of the distinctly *non-diverse* group of historians who have traditionally interpreted them.

Research into the PCnet must be part of the "nuclear history renaissance." This will require new research methods and new ways of thinking about the bomb. For example, to find out if — and if so, how — the first "atomic admiral" influenced the design of the proving ground that would test the world's first atomic bomb, it will require 1) breaking the code on newly declassified documents; 2) reinterpreting previously examined documents; 3) incorporating the fresh perspective of new and diverse researchers and scholars.

Christman was partially right to say Vogel's theory of a nuclear explosion was based on "Parsons' appearance at the site, plus the report of a mushroom cloud." Those are indeed valuable research leads, clues to an even larger body of evidence. For example, the known hazards at Port Chicago, which could have been avoided or addressed, support the claim that Port Chicago was designed to be the world's first, top-secret nuclear proving ground.

Improbable cause

It took but a little sleight of the Navy's hand to turn the story of the Port Chicago disaster into the story of the Port Chicago mutiny.

The cause of the explosion has never been determined. Perhaps, though, a fresh examination with twenty-first century technology could yield new information. Or at least the old history can be reviewed, using new evidence to address the question of whether there may have been a nuclear test at Port Chicago.

The shifty answer to that question on the **Port Chicago Naval National Memorial** (POCH) National Park Service website is a veiled acknowledgement that other than the COI investigation, there has been no effort to determine the cause. After all, once the Court of Inquiry said it never would be determined, there was no further need for investigation. The case rests – on the slender conclusion that a nuclear test was "unlikely."

The NPS explanation perpetuates the common and convenient *assumption* that the blast was accidental.

> "Remember that WW II was going full tilt in 1944. With no end in sight, we were in it for the long haul. Port Chicago was operating 24 hours a day, 7 days a week, with crews loading munitions continually. That, and unfortunately munitions loading was being done by men with little or no training. Men were also told the munitions they were loading was not active and would be armed upon arrival to the Pacific theatre. Other factors include supposed betting by officers to see whose crew could load the most munitions the fastest during their shift. **It is *unlikely* that enemies of the United States managed to sabotage Port Chicago, or, as *some claim, that the U.S. was testing a nuclear bomb*.** All it really took was an errant shell dropping to the deck from a cargo net and...5,000 tons of munitions went off."[109] [emphasis added]

Unlikely? Unsavory. Unpleasant. Unproven. ***Uninvestigated.***

Blame without a cause

Most accounts emphasize the poor training of the Black sailors, who worked under pressure, supervised by White officers, including some who were inexperienced, some who were racially biased, and some who were both. (Inexperienced does may refer to career experience, but the officers not only had little training in handling ammunition, they also had insufficient training in **The Supervision of Negro Personnel**.)[110]

The claim that the Black sailors were of "low quality" has become embedded in the history of Port Chicago[111]; but it was an exaggeration by White officers who were opposed to the recent integration of the Navy. Examination of their Bluejacket personnel files reflect a variety of experience among the Port Chicago sailors.

Why did the Navy choose to staff this vital facility with sailors who had the lowest scores in boot camp? And what did it even mean to have a score low on a test? A test of what? Ammunition handling was not one of the things the recruits were taught at the Great Lakes Naval Training Center. Commercial stevedores underwent extensive training before handling high explosives, yet the Navy refused their offer to train Port Chicago personnel. Why?

Given the emphasis on race, historians have overlooked several implications of the administrative decisions—including the fact that "rough handling by an individual or individuals", which could have occurred "at any stage of the loading process" is the only second of six probable causes of "the initial explosion."[112]

The third, fourth and fifth probable causes may contain more clues to the PCnet; or they may be distracting filler. But the omission of one quite probable cause may be evidence of a cover-up.

The Court officially stated that no one was at fault. That contradicts the common wisdom that says the Black sailors were blamed. However, the disparaging language about the quality of personnel, combined with the court martial, creates a convenient scapegoat. The impression then backs up the conclusion that the explosion was an accident. With the Court's conclusion that the cause could never be determined, further investigation was unnecessary. Case dismissed, assumptions intact.

Until new evidence—like the report of a plane dropping a bomb near Mare Island just before the explosion—reopens the case.

51. That the probable causes of the initial explosion listed in the order of probability are:

a. Presence of a supersensitive element which was detonated in the course of handling.

b. Rough handling by an individual or individuals. This may have occurred at any stage of the loading process from the breaking out of the cars to final stowage in the holds.

c. Failure of handling gear, such as the falling of a boom, failure of a block or hook, parting of a whip, etc.

d. Collision of the switch engine with an explosive loaded car, possibly in the process of unloading.

e. An accident incident to the carrying away of the mooring lines of the QUINAULT VICTORY or the bollards to which the QUINAULT VICTORY was moored, resulting in damage to an explosive component.

f. The result of an act of sabotage. Although there is no evidence to support a sabotage as a probable cause, it cannot be ignored as a possibility.

"*Presence of a supersensitive element which was detonated in the course of handling*" was first among the six most probable causes of the Port Chicago explosion identified by the Court of Inquiry.[113]

"A plane dropping a live bomb"

Sabotage was unlikely. The COI determined that there was "no evidence to support sabotage as a probable cause" of the Port Chicago explosion.

This suggests that the three members of the Court (which, again, included Parsons' brother-in-law, Captain James Crenshaw) either ignored or did not have access to reports from Mare Island (where Crenshaw may have been stationed).

Two documents from the July 1944 War Diary provide sufficient reason to investigate the possibility of an enemy attack.

At 10:25 P.M on July 17th, an officer who resided in the area, identified only as Lieutenant Snedeker, "reported a plane dropping a live bomb near his home." For whatever reason, the COI makes no mention of this report. Snedeker's name does not appear on the witness list.

> 17 July -- At 2225, Lieutenant Snedeker reported a plane dropping a live bomb near his home. The explosion "nearly blew him out of bed". At 2250, a second explosion was reported by the Tower. A local broadcasting station reported an explosion at Port Chicago.

"July 17—At 2225, Lieutenant Snedeker reported a plane dropping a live bomb near his home. The explosion "nearly blew him out of bed". At 2250, a second explosion was reported by the Tower. A local broadcasting station reported an explosion at Port Chicago."[114]

The report from Lt. Snedeker is recorded in the Mare Island War Diary, July 17, 1944. This significant public record has been overlooked in histories of the Port Chicago disaster. The report that a plane was dropped from a bomb in the vicinity of Port Chicago around the time of the explosion becomes of even greater interest in the light of other evidence, including two points that *were* included in the COI report.

The six most probable causes are listed in order of likelihood. First on the list is "presence of a supersensitive element which was detonated in the course of handling," a condition that might have occurred in a depth bomb.

The term "supersensitive component" as herein used is defined as:

a. One wherein a thin film of high explosives is present because of defects in the manufacture of the case or faulty filling of that particular component. (This condition could have occurred in the Mark 47 and the Mark 54 depth bombs.)

Mark 54 "depth bomb" cited as example of "supersensitive component"

An Army Air Force pilot happened to witness the explosion. In his testimony to the COI, he described it as being similar to the explosion of a depth bomb:

> "...The explosion itself, if I would compare it to something I have seen before, I would compare it to the sight when you see a depth bomb explode in the water. The explosion went outward, as a depth bomb does in the water, and then the surge of the explosion went up like the water does after the depth bomb explodes. That is about the nearest I could come to accurately describing it."

Note that observation planes accompanied the B-29s that dropped the bombs on Hiroshima and Nagasaki, and flew above all of the early bomb tests as well. Parsons was part of the team that observed the Trinity test.

In a September 1944 letter to Groves, Parsons used the phrase, "From my observation of the Port Chicago explosion . . ." Prior to his trip to the West Coast, he enlisted the help of the Fourth Air Force to support experiments "on the West Coast".

It seems reasonable that the report of a plane dropping a bomb (or a depth charge) near Port Chicago around the time of the explosion would have been cause for high alert. The omission of this fact from the COI investigation, and from the history of Port Chicago, requires an explanation.

These two obscure facts show the logistical possibility that a secret test could have been conducted at Port Chicago. This lends credence to the PCnet, but it would require further investigation to establish for a fact that the explosion at the accident-prone naval magazine resulted from an intentional detonation.

Coincidences? or coordinated incidents?

As a security measure, General Leslie R. Groves, Director of the Manhattan Project, enforced the military policy known as "compartmentalization." To prevent security leaks, information was shared only on a "need-to-know" basis.

History tends to be written from a vertical view, presenting events as if they took place only in sequential, chronological order. This in effect "compartmentalizes" historical events. It requires a different perspective to recognize possible correlations between events that have been reported as separate and distinct. The PCnet provides a horizontal viewpoint, revealing significant links between several sets of overlapping historical events.

For example, the Trinity test of July 16, 1945 shares an anniversary with the July 17, 1944 explosion Port Chicago explosion. That might just be an uncanny coincidence; but it could also be evidence of careful planning. In either case, the false press release issued by General Groves after the Trinity test not only links the two incidents, it also demonstrates that an explosion at a remote ammunition magazine made a good cover for a nuclear test.[115]

NO ONE INJURED IN AMMUNITION BLAST

ALAMOGORDO, N. M., July 16.—(AP)—An ammunition magazine on the Alamogordo air base reservation exploded with such force it was seen for 100 miles and heard as far away as eastern Arizona.

Col. William O. Eareckson, commanding officer, said the magazine was remotely located and that no one was killed or injured by the blast. He added that "weather conditions affecting the content of gas shells exploded by the blast may make it desirable for the army to evacuate temporarily a few civilians from their homes."

On July 16, 1945, one year after the Port Chicago explosion of July 17, 1944, General L. R. Groves used the false report of an explosion at a remote ammunition depot as a cover for the Trinity test.

During the week of July 17, 1944, several historic coincidences took place that seem to have escaped the attention of journalists and historians. Some of the events may have gone unreported because of the voluntary self-censorship of the press during the war. Or it may be that some events did not seem newsworthy at the time, especially since war news in the headlines included the downfall of Japanese Warlord Hideki Tojo and the attempted assassination of Adolf Hitler.

The Bethlehem Globe-Times
COAST EXPLOSIONS KILL HUNDREDS
Tojo Relieved As Chief Of Jap Imperial Army Staff
American Forces Capture German Fortress Of St. Lo
2 Navy Ammunition Ships Explode At Port Chicago, California, With Deaths Listed From 200 To 650

There was a lot going on in the world during the week of the Port Chicago explosion. Front page, Bethlehem Globe-Times, July 18, 1944

• ***July 14th*****:** After announcing his candidacy for a fourth term on the 11th, FDR slipped out of DC on the 13th, stopping overnight in New York before commencing a secret cross-country train trip to the West Coast.

• ***July 14th****:* Admiral Carleton E. Wright, Commandant of the Twelfth Naval District, issued the *Precept for General Court Martial,* the official document authorizing the Port Chicago Mutiny trial.

• **July 15th:** FDR stopped in Chicago for a brief meeting with Democratic Party, informing them of his choice of Senator Harry Truman for a running mate. (Various versions of the story add intrigue. These include reports of the blatant political shenanigans that led to Truman's upset victory over Vice-President Henry Wallace.)

• ***July 17th***: FDR's flagship departed Mare Island Navy Shipyard, accompanied by the entire COMINCH fleet, which had been anchored there since July 6th.

• ***July 17th***: A massive explosion at the remote Port Chicago Naval Ammunition Depot near San Francisco destroyed two ships, killing 320 people. According to a local newspaper, most victims were "atomized".

• ***July 17th***: Manhattan Project leaders meeting at the University of Chicago decided to shelve the *Thin Man* plutonium gun-assembly bomb, the world's first atomic bomb model. The decision led to the reorganization of the Los Alamos laboratory and a delay in the scheduled delivery of the first combat-ready nuclear weapon.

• **July 19th**: The Democratic National Convention got underway in Chicago, Illinois. Party bosses took drastic measures to assure the upset victory of Senator Harry Truman over Vice President Henry Wallace.

• **July 19th**: FDR arrived in San Diego in the early hours of the morning. His train, the *Ferdinand Magellan* and his flagship, the *USS Baltimore* (CA-68), appear to have arrived in San Diego on the same day.

• **July 20th**: FDR suffered a severe bout of pain that left him writhing on the floor of his train car. He recovered without consulting his doctor, who was traveling with him, and went on to observe a marine landing operation. In a letter to first Lady

Eleanor Roosevelt, he later downplayed the incident, describing it as a case of "the collywobbles". For some reason, he said the attack occurred after he observed the military exercise, not before.

- **July 20th**: FDR accepted a fourth term. He announced that he was delivering his acceptance speech "from a naval base on the West Coast", but did not mention the West Coast Naval base that had just made headline news, the worst homefront disaster of World War II.

- **July 21st:** Senator Harry Truman was nominated for Vice-President. Despite FDR's failing health, only a handful of people seemed to foresee that the "accidental president" would inherit two enduring items on FDR's global agenda, the atomic bomb and the United Nations.

- **July 21st:** FDR boarded the *USS Baltimore* CA-68 for the cruise to the Honolulu conference.

These remarkable and coincidental events may appear to be unrelated, except as they shed light on the Port Chicago nuclear explosion theory – which, in turn, brings new perspective to the gap-ridden story of the bomb.

Discredit? or distraction?

Of all the PCnet-related events that took place during the week of the Port Chicago explosion, perhaps the most puzzling was the issuance of the "Precept for General Court Martial" on **July 14th, 1944**, three days *before* the explosion and weeks before 328 Black sailors refused to load ammunition onto ships, citing fear of another explosion.

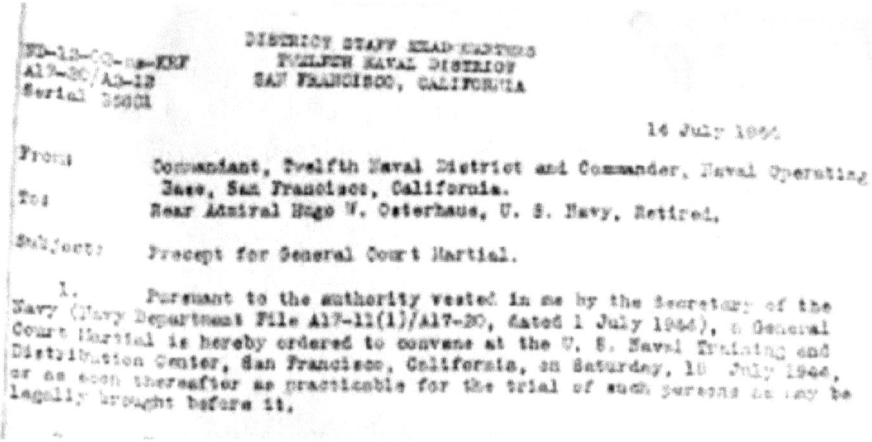

DISTRICT STAFF HEADQUARTERS
TWELFTH NAVAL DISTRICT
SAN FRANCISCO, CALIFORNIA

A17-20/A3-13
Serial

14 July 1944

From: Commandant, Twelfth Naval District and Commander, Naval Operating Base, San Francisco, California.
To: Rear Admiral Hugo W. Osterhaus, U. S. Navy, Retired.

Subject: Precept for General Court Martial.

1. Pursuant to the authority vested in me by the Secretary of the Navy (Navy Department File A17-11(1)/A17-20, dated 1 July 1944), a General Court Martial is hereby ordered to convene at the U. S. Naval Training and Distribution Center, San Francisco, California, on Saturday, 15 July 1944, or as soon thereafter as practicable for the trial of such persons as may be legally brought before it.

The PCnet requires an explanation for why the Precept authorizing the Port Chicago mutiny court martial was issued on July 14, 1944—three days before the massive explosion that led to the "work stoppage".

In my brief exchange with Vogel, he referred to some earlier discussion of a possible court martial. He was not more specific about any violations or perpetrators, merely suggested that the Precept may have originally been created with the intention of conducting a different court martial. That possibility warrants further investigation. However, my cursory survey of records from the Twelfth Naval District showed no subsequent trial that was authorized by the same Precept.

Vogel may have been referring to criminal offenses that took place at the base in 1943, as described by James Campbell, in **The Color of War**, who writes about talk of "mutiny" as early as July 1943.[116]

Lieutenant Raymond Robert "Bob" White was a White officer who apparently had rapport with the Black sailors. He told his wife, Inez, about a drowning that had taken place. In a letter to her in-laws, she used the word "mutiny", explaining that the "colored boys" had refused to work because they felt that one of the officers should have saved the sailor who drowned.

According to Campbell, this was not the first slowdown or work stoppage. Another officer, Lieutenant Lee Cordiner, had resorted to extreme discipline: those who refused to work would not eat. But hungry sailors pushed right past him and entered the mess hall, some threatening his life. Captain Nelson Goss, the high-profile commander of the Mare Island Navy Shipyard, ordered a deck court (possibly, Campbell speculates, to *avoid* the notoriety of a mutiny trial).

As noted, there were ongoing hazards at the base, as well as increasing pressure to load more ammunition faster. Additionally, some of the Black sailors felt they were being treated unjustly on account of their race. In boot camp, they had trained for all types of jobs, but most Black seamen were assigned to shore duty, to labor jobs like loading ammunition. Racial tensions increased. One White officer apparently committed suicide. There was a fight over a card game; someone got shot. To put it mildly, morale at the base was dismal.

That said, it does not follow that any or all of these circumstances led Wright to issue the Precept for the Port Chicago Court martial dated July 14th, 1944, with the urgent call to convene the next day, a Saturday.

Even if the work stoppage was inevitable, a general court martial – the Navy's most extreme form of discipline, which carried the possibility of the death penalty – was not.

The memorandum, issued on a Friday, calls for establishment of a proceeding the very next day, Saturday, July 15, 1944, "or as soon thereafter as is practicable" for the trial of unspecified persons. It appears that the only proceeding authorized by this document was the Port Chicago mutiny trial.

It is possible that another planned proceeding was disrupted by the turn of events, but the urgency of the language rules against that likelihood. By the same token, though, the simplest explanation for the early date of the document is almost unthinkable.

No single fact demands an investigation of the PCnet more than the idea that the trial may have been pre-planned as a red herring to distract attention from the massive explosion. The PCnet is considered a conspiracy theory, and this may sound like just another outlandish notion, but it is based on the evidence: **the date requires explanation**. Only an investigation can rule out the possibility that such an outrageous plan was actually carried out – especially since the mass court martial did indeed serve that very purpose. The trial not only upstaged the blast, immediately and permanently, it became the defining event of the disaster.

Two counterfactual ideas – alternative histories (or, 'what if's') – can help us appreciate the severity and significance of this concern. First, what if there had been no explosion? Even if there had been a work stoppage, would it have resulted in a mass trial, a general court martial that carried the death penalty? Secondly, what if there had been no court martial? Would the Port Chicago explosion, "the worst homefront disaster of World War II", still be a little-known event in African-American history?

Many people believe the 'work stoppage' was a protest; that premise helps explain how the injustice of the trial led to the integration of the Navy. Robert Allen conveys this perception in **The Port Chicago Mutiny**, the seminal history on which most other accounts are based.

Noting the sailors' ongoing discontent with conditions at the base, Allen (who studied the event for his PhD in sociology) concluded that the "minor work slowdowns and stoppages had occurred in the past" were "precursors of the work stoppage that occurred after the explosion.[117]

> That work stoppage was inevitable. It would have happened. But something else had to happen to give it a shove. The explosion was the instrument by which all of this injustice was brought to

> light. Had the explosion not happened, 320 men would not have lost their lives, but eventually something would have happened to bring about this work stoppage — that the conditions might be exposed.

Still, as Allen reviewed the data, he was

> . . . struck by the apparent lack of significant acts of mass resistance before the explosion. The enlisted men certainly had serious grievances and other matters about which they had long been disturbed, but aside from some individual acts of defiance and one brief work stoppage, there was no evidence of any previous collective action. Why?

Allen found "no change in the men's military status before and after the explosion" that would explain the timing. The sociologist applied the skills of his trade to develop the elaborate and scholarly construct that has become the widely accepted basis for the men's collective action. Something, Allen realized, must have changed.

> . . . there must not only have been motivating "grievances," but there must also have been a breakdown in old modes of thinking and behaving which had inhibited collective action to redress the grievances.

But when attempting to determine what might have led the men to think and act differently, Allen put little store by claims that *outside agitators had instigated the action*. After all, that charge was made by Captain Nelson Goss, described by Campbell as someone "who may have been more than a garden-variety bigot."

To Goss, ". . . the source of the problem was "agitators, ringleaders, among these men." However biased Goss may have been, the question for the PCnet is, was that perception due *only* to the officer's obvious racial prejudice? Or was there evidence to corroborate his observation? After all, his racist attitudes could have easily co-existed with the possibility that the men were "activated" by something other than either anger or "mass fear" (as FDR explained).

Indeed, during the court martial at least one witness testified about leaders who argued in favor of the work stoppage but then backed out. From the perspective of the PCnet, that *could* be seen as the clever tactic of an effective provocateur.

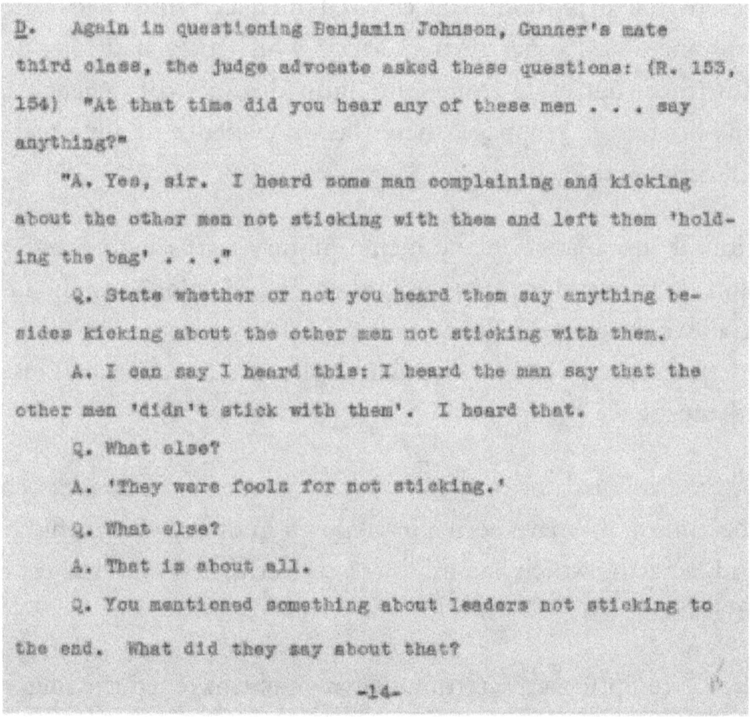

D. Again in questioning Benjamin Johnson, Gunner's mate third class, the judge advocate asked these questions: (R. 153, 154) "At that time did you hear any of these men . . . say anything?"

"A. Yes, sir. I heard some man complaining and kicking about the other men not sticking with them and left them 'holding the bag' . . ."

Q. State whether or not you heard them say anything besides kicking about the other men not sticking with them.

A. I can say I heard this: I heard the man say that the other men 'didn't stick with them'. I heard that.

Q. What else?

A. 'They were fools for not sticking.'

Q. What else?

A. That is about all.

Q. You mentioned something about leaders not sticking to the end. What did they say about that?

-14-

During the Port Chicago court martial, Gunner's Mate Benjamin Johnson testified about "leaders not sticking to the end" – possibly corroborating the idea that there had been agitators involved. That would explain how someone could have predicted the need for a court martial three days before the explosion.

Further research is warranted. To be clear, there is a difference between *claiming* that the trial was designed to be a decoy and *proving* that that was the case. What is necessary here, is to *investigate* the reason for the early date on the *Precept for General Court Martial.* We can make no valid assumptions whatsoever. Perhaps the document was intended for a scheduled proceeding in the case of a recent event that was more severe than those that had *not* warranted a court martial months before.

Perhaps there was a build-up of events, and a case was about to come to trial but was set aside because of the explosion and the events that followed. If so, perhaps a careful investigation would yield documentary evidence of that planned hearing, and thus satisfy the question. Without that research, though, the *unexamined evidence* bears *potentially* sinister connotations. One thing is clear: the Precept issued on July 14th was the official document that authorized the Port Chicago trial.

It is also clear that the Port Chicago mutiny trial was an effective and enduring distraction from the Port Chicago explosion. This was no ordinary court martial; it was the largest tribunal in U.S. naval history. The massive blast – that killed 320 people, destroyed two ships, made international headlines, is still counted among the greatest man-made explosions in history, and *provided data vital to the research and development of the atomic bomb* – was soon upstaged by the mass trial.

The public is usually barred from military trials, yet the Navy itself publicized the Port Chicago case and welcomed the participation of a civilian attorney. Thurgood Marshall, who later became the first African American Supreme Court Justice, was lead attorney for the NAACP. The involvement of the civil rights organization further publicized the mutiny trial, which became *the* story of the Port Chicago disaster.

Marshall said the trial was a frame-up, "deliberately planned and staged by certain officers to discredit Negro seamen." That may have been a reasonable conclusion, judging by some of the racist language in documents where Naval officials discussed the character and morale of Negro sailors, language that has become embedded in historical accounts of the Port Chicago disaster.[118] On the other hand, it would have taken little more than a bad conduct discharge in their Bluejacket files to discredit the sailors.

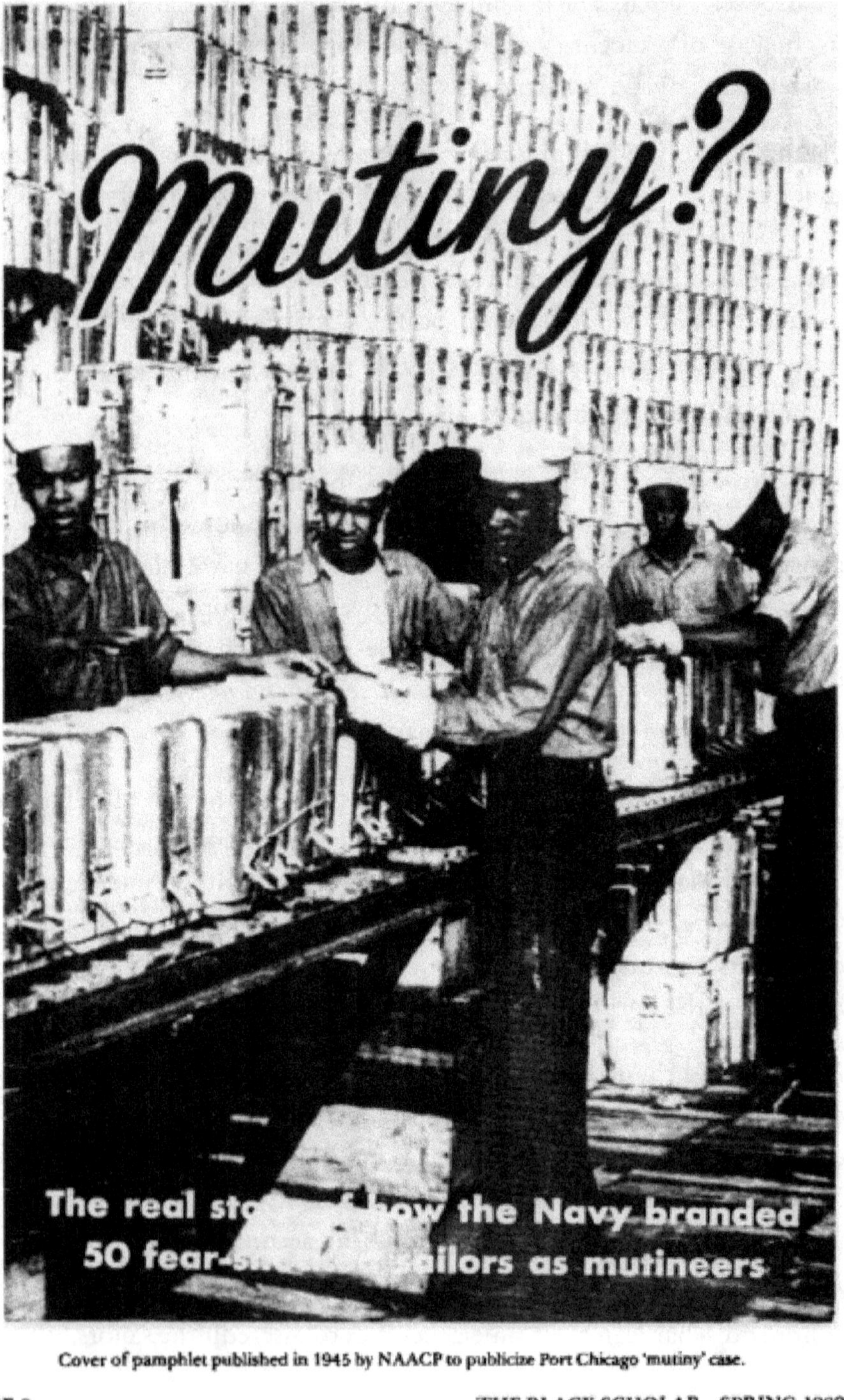

Cover of pamphlet published in 1945 by NAACP to publicize Port Chicago 'mutiny' case.

The NAACP issued a pamphlet to publicize the Port Chicago court martial. Involvement of the NAACP and publicity by the Navy brought unusual public attention to the military tribunal, which upstaged the story of the explosion itself.

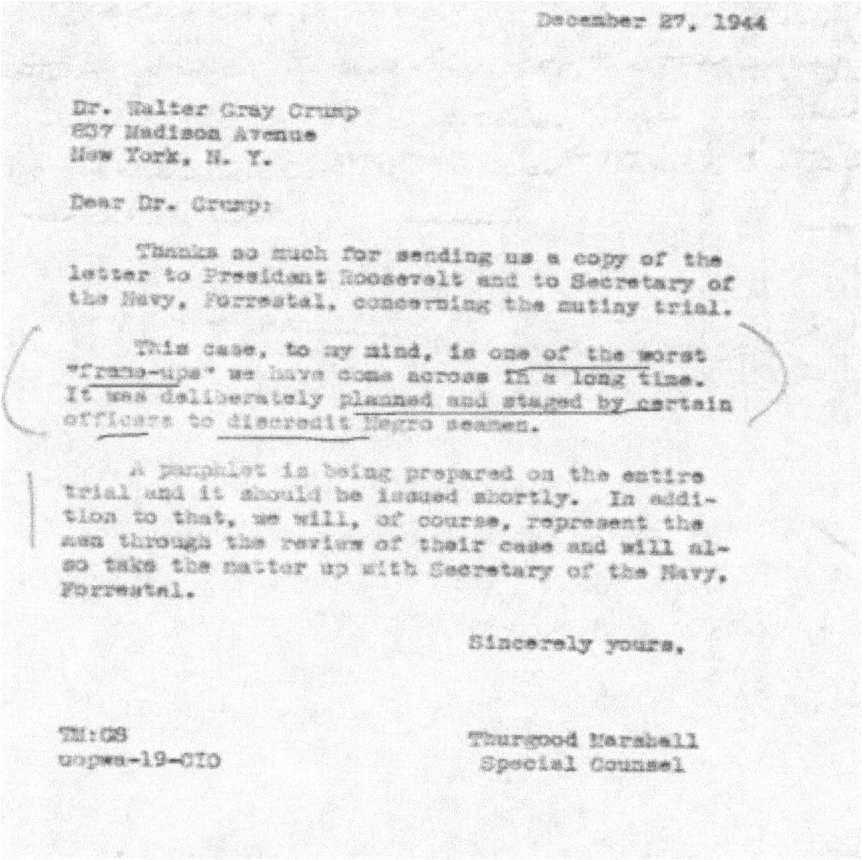

December 27, 1944

Dr. Walter Gray Crump
807 Madison Avenue
New York, N. Y.

Dear Dr. Crump:

Thanks so much for sending us a copy of the letter to President Roosevelt and to Secretary of the Navy, Forrestal, concerning the mutiny trial.

This case, to my mind, is one of the worst "frame-ups" we have come across in a long time. It was deliberately planned and staged by certain officers to discredit Negro seamen.

A pamphlet is being prepared on the entire trial and it should be issued shortly. In addition to that, we will, of course, represent the men through the review of their case and will also take the matter up with Secretary of the Navy, Forrestal.

Sincerely yours,

TM:GS
uopwa-19-CIO

Thurgood Marshall
Special Counsel

NAACP attorney Thurgood Marshall (the first African American to become a Supreme Court Justice) thought the Port Chicago mutiny trial was a "'frame-up' . . . to discredit the Negro seamen." But what if there was another reason to frame them?

After Wright threatened to invoke the death penalty, some 258 sailors relented and went back to work. They were subject to summary court martial, a milder form of discipline than a general court martial faced by the Port Chicago 50. When their case came across President Roosevelt's desk, the compassionate Commander-in-Chief of the Army and Navy —who did like a good play on words — recommended leniency. After all, he said, "they were *activated* by *mass* fear."

Perhaps he knew from personal experience.

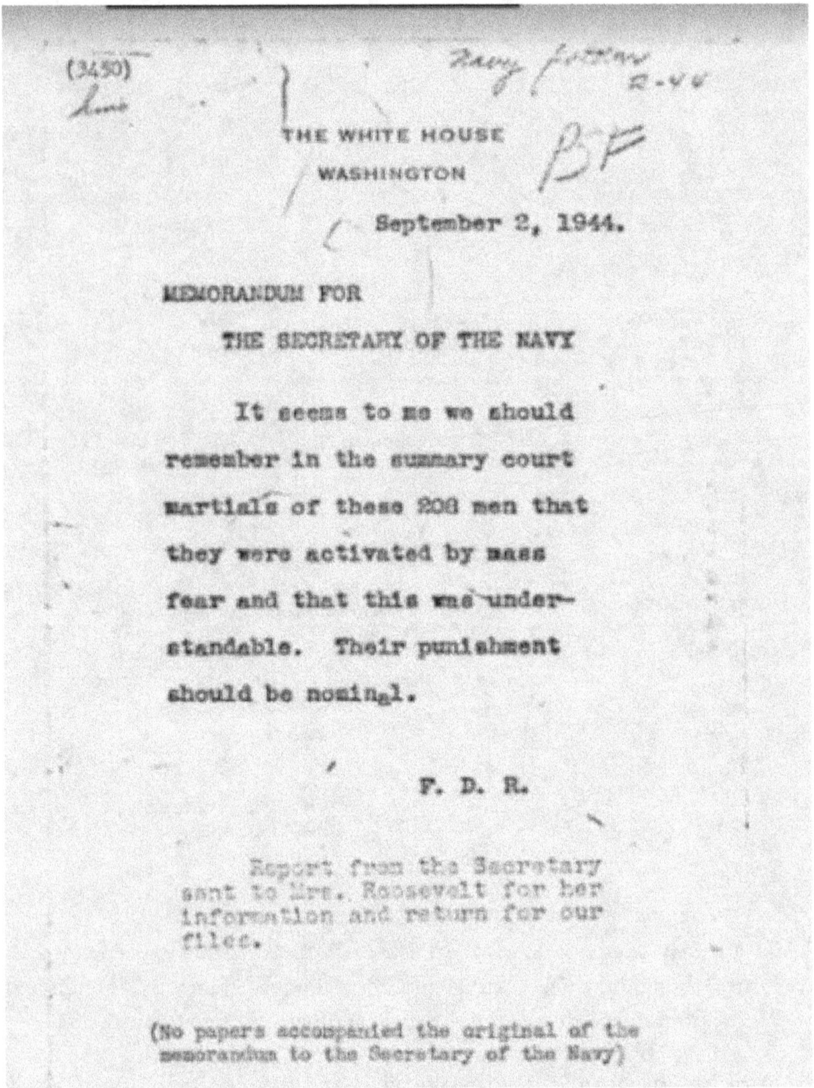

THE WHITE HOUSE
WASHINGTON

September 2, 1944.

MEMORANDUM FOR

THE SECRETARY OF THE NAVY

It seems to me we should remember in the summary court martials of these 208 men that they were activated by mass fear and that this was understandable. Their punishment should be nominal.

F. D. R.

Report from the Secretary sent to Mrs. Roosevelt for her information and return for our files.

(No papers accompanied the original of the memorandum to the Secretary of the Navy)

FDR, Commander-in-Chief of the U.S. Army and Navy, recommended leniency in the summary court martial of the sailors who returned to work because ". . . **they were activated by mass fear**." Through the lens of the PCnet, this unusual choice of words may have been one of FDR's characteristic puns. [emphasis added] - FDR to James Forrestal, Secretary of the Navy, September 2, 1944

On his way back to Washington after his July 1944 trip to the Honolulu Conference, FDR made a rambling, discordant speech from the Puget Sound Naval Shipyard at Bremerton, Washington. He reported how he had watched a Marine landing operation "from a high bluff above the sea."

The phrase could apply, as well, to a convenient spot (say at Bernicia State Park, near Mare Island) where it would have been possible to observe the world's first nuclear blast from a safe distance.

Considering the effect of the rigged court martial, it is reasonable to speculate that a deliberate plan to achieve a greater purpose may have warranted the mass trial of 50 sailors.

Focus and perspective

Recapping key facts:

Admiral Carleton H. Wright, Commander of the Twelfth Naval District, authorized the Port Chicago court martial in a memo dated July 14, 1944.

When he spoke with Wright at the Mare Island Navy Shipyard on July 20th, Captain William S. Parsons, head of the ordnance division of the Manhattan Project, was en route to Port Chicago, to study 'the effects of the detonation." Parsons was accompanied by his brother-in-law, Captain Jack Crenshaw, a member of the three-man Court of Inquiry that investigated the explosion.

Weeks later, in August, when 328 Black sailors refused to return to the hazardous duty loading ammunition onto ships, Wright was the officer who threatened to make sure they got the death penalty.

That fall, fifty of the sailors who refused to load ammunition onto ships were found guilty of conspiracy to commit mutiny – in a trial that appears to have been planned before the explosion that led to their "work stoppage."

In his book, **The Port Chicago Mutiny**, Robert Allen defined the sailors' action as a "work stoppage". The concept carries the idea that the men were protesting racism, though, as Allen noted, the Navy makes no allowances for such "civil disobedience". The sticky phrase undergirds the popular notion that *the* Port Chicago story was about racial injustice.

As Allen reported in **The Black Scholar** (Winter 1994), a 1994 Navy review of the Port Chicago mutiny trial found that discrimination was indeed a factor at Port Chicago. Yet the Board for Correction of Naval Records refused to overturn the verdict, concluding that "neither racial prejudice nor other improper factors tainted any part of the Port Chicago court martial proceedings."[119]

The irony is that discrimination was at most a minor issue in the actual trial. The defense was based on safety issues: the men did not refuse to obey all orders; rather, their refusal to load ammunition onto ships resulted from fear of another explosion.

Allen, the leading expert on the Port Chicago disaster, interviewed some of the men from the "Port Chicago 50," and wrote his book from their point of view. The emphasis on racism may have been theirs, or his, or both. In either case, his book minimizes their claim that they were afraid to load ammunition onto ships after the explosion. In the end, the focus on race and injustice helped shape the narrative, as did the early involvement of the NAACP.

As noted, Allen broached the question of why the Black sailors at Port Chicago did not stage the work stoppage before the explosion. The question is approached differently from the PCnet point of view, which focuses on the explosion rather than the court martial.

The critical point to consider is how the focus of the narrative developed, and why. Is it possible that the very real racial issues at Port Chicago were *exploited* in order to cover up a secret nuclear test?

In addition to the safety hazards at Port Chicago, racial injustice was a genuine grievance. As Allen knew, some of the sailors felt they had been tricked into joining up; they sought a "new deal." Still, other than the hostile remarks of Admiral Wright, nothing in the record discounted their claim that they refused to return to the docks because they were afraid.

In the same way other aspects of the history conceal facts related to the PCnet, the theme of racial injustice overshadows the fact that the explosion at Port Chicago was no ordinary blast. Its force was studied by the men who were designing the atomic bomb.

Although there was no attempt, violent or otherwise, to usurp authority, the Navy chose to conduct a mass trial of 50 sailors, trying them on a controversial charge of "conspiracy to commit mutiny", rather than summarily disciplining each man for failure to obey orders. Why the overkill? Far more serious infractions at other military bases during the war met with far less stringent responses.

To explain why the Navy took unusual measures to publicize the military tribunal, historians make ungrounded assumptions. Some say, for example, that the Navy wanted to make sure the public knew the sailors were being treated fairly – as if Naval justice was a civilian concern and the Naval court answered to the public. Others speculated that the Navy wanted to caution other potential protestors that violations would meet with severe justice – as if the **Bluejacket Manual** did not already spell out the consequences of violations.

The PCnet shows that (and how) racism – a real condition at Port Chicago – could have been exploited to misdirect attention from the explosion. Goss' noted racial bias may have made him prone to interpret the Negro sailors' assertiveness as aggressiveness, and his prejudices may have influenced his conclusion that there were agitators among the group; but that does not necessarily mean he was wrong.

The July 14, 1944 memo authorizing the Port Chicago court martial called for a Court to convene the very next day, Friday July 15th, "or as soon thereafter as practicable" "for the trial of such persons as may be legally brought before it." There is no indication of who had violated what infraction. Barring some other less obvious explanation, the date suggests that someone knew in advance that a violation was going to take place, and soon. How would they know that?

Whatever the charge was to be, the early date indicates either that 1) an unspecified act that had already taken place was to be taken up by a court on an urgent basis, or 2) someone had foreknowledge of some unspecified violation that would soon occur.

In the first case, the questions would be 1) why the violation and accused perpetrators were not specified, and 2) why it appears that no court martial for that apparently serious offense was ever held. In the latter case, the question would be how anyone could have predicted that "such persons" might do something to warrant a court martial.

As Campbell noted, there had been earlier talk of a court martial earlier. It is possible that those grievous incidents came to a head on July 14th – and that that date just happened to coincide with the day FDR began his secret trip to the West Coast.

These distinct yet co-occurring events share at least three elements: the date, July 14, 1944; the West Coast location; and presence of FDR's flagship at the Mare Island Navy Ship. The overlap is seen only through investigation of the PCnet. In fact, the location of FDR's flagship prior to the explosion, a topic that might seem to be of general public interest, especially since the President's ship departed on the day of the explosion, does not show up in any historical account until the history is viewed from the perspective of the PCnet.

Comparable effects

The effects of the explosion at Port Chicago were in some ways comparable to those of the bomb that destroyed Hiroshima. Most specifically, the obliteration of the *SS E.A. Bryan* and the obscure news report that most victims were "atomized" evokes pictures of shadows representing those vaporized in the attack on Japan.[120]

This composite picture compares scenes from Hiroshima and Port Chicago. **In both events, many of the victims were vaporized, or "atomized."** Further research would be necessary to establish the similarities and differences between the effects of the two unprecedented blasts. [121]

On July 17th, 1944, Manhattan Project leaders meeting at the University of Chicago decide to shelve the *Thin Man* bomb. Parsons, head of the Ordnance division, the man responsible for production and delivery of the bomb, was absent from the meeting. He had worked on the prototype plutonium gun-assembly bomb for over a year. Why was he not present to explain the status of the bomb and contribute to future plans?

On that same date, July 17th, 1944, a massive explosion at the Port Chicago Naval Ammunition depot near San Francisco, obliterated one ship and broke another in half. The next day, a local newspaper reported that most of the 320 victims were "atomized."

These are not typical effects of a conventional explosion. In fact, there was no similar destruction to any of the ships that were damaged in the grand test of atomic bombs against naval ships at *Operation Crossroads* in July 1946.

On the other hand, the obliteration of the *E.A. Bryan* and the "atomization" of victims at Port Chicago is comparable to the "vaporization" of victims of the atomic bomb attacks on Hiroshima and Nagasaki.

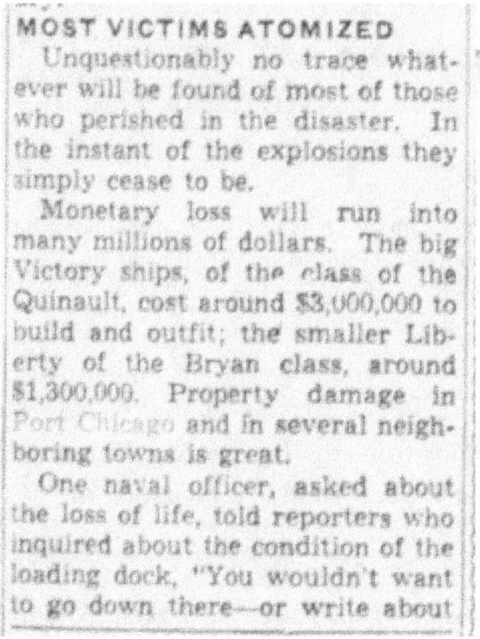

MOST VICTIMS ATOMIZED

Unquestionably no trace whatever will be found of most of those who perished in the disaster. In the instant of the explosions they simply cease to be.

Monetary loss will run into many millions of dollars. The big Victory ships, of the class of the Quinault, cost around $3,000,000 to build and outfit; the smaller Liberty of the Bryan class, around $1,300,000. Property damage in Port Chicago and in several neighboring towns is great.

One naval officer, asked about the loss of life, told reporters who inquired about the condition of the loading dock, "You wouldn't want to go down there—or write about

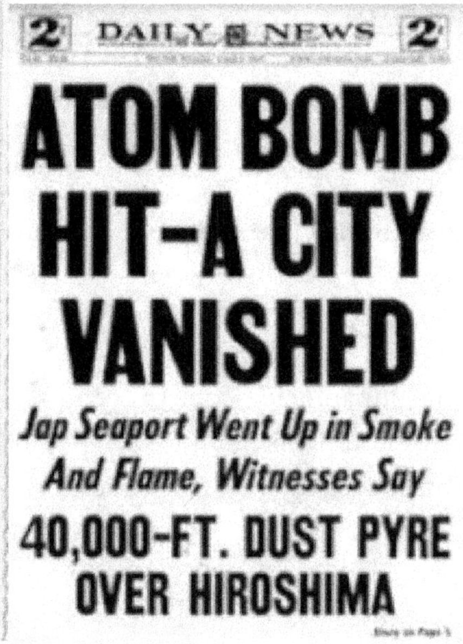

2 DAILY NEWS 2

ATOM BOMB HIT-A CITY VANISHED

Jap Seaport Went Up in Smoke And Flame, Witnesses Say

40,000-FT. DUST PYRE OVER HIROSHIMA

Port Chicago: "Atomized"

Hiroshima: "Vanished"

"Atomized" . . . "Vanished" – One clue to the PCnet is found by comparing this detail from a small clipping from a local San Francisco newspaper report about the Port Chicago explosion with national headlines about the atomic bomb attack on Hiroshima. **To adequately analyze the unusual effect, further research of the PCnet is necessary.** (The effects of nuclear explosions vary with the type of test, material used, type of device, height of detonation, weather and climate, and many other factors. The tower used in the Trinity test vanished in the explosion, but no such effect was reported after *Operation Crossroads*.)

The PCnet and democracy

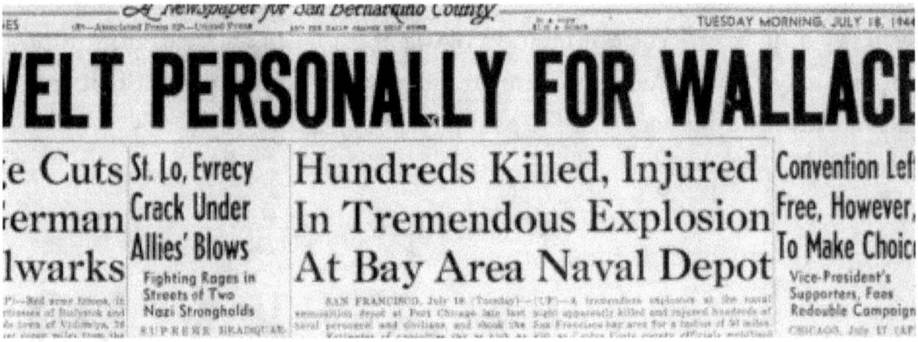

A Newspaper for San Bernardino County

TUESDAY MORNING, JULY 18, 1944

VELT PERSONALLY FOR WALLACE

e Cuts erman lwarks

St. Lo, Evrecy Crack Under Allies' Blows

Fighting Rages in Streets of Two Nazi Strongholds

Hundreds Killed, Injured In Tremendous Explosion At Bay Area Naval Depot

Convention Lef Free, However, To Make Choic

Vice-President's Supporters, Foes Redouble Campaign

The Port Chicago explosion shared headlines with the Democratic National Convention where FDR received his fourth nomination. According to bold headlines in the *San Bernardino Sun*, FDR supported Vice-President Henry Wallace.

On July 20th, 1944, FDR accepted the nomination for an unprecedented and unparalleled fourth term as President of the United States. He broadcasted his acceptance speech not from the floor of the Democratic National Convention in Chicago, Illinois, but "from a Navy base on the West Coast."

On July 21st, 1944, Democratic Party bosses shut down their National Convention in Chicago, Illinois, derailing Vice-President Henry Wallace's likely nomination for another term on the Roosevelt team. The next day, after some bold political maneuvering, Senator Harry Truman, an (alleged) underdog who'd (reportedly) once protested that he wouldn't be President for a million dollars, won an upset.

The whole scene was staged, both in time and in history. FDR claimed on one occasion that he barely knew the man who headed the powerful Truman committee established by the Senator himself in 1941 to monitor the FDR administration's (public, legal) wartime defense spending. The popular two-term senator was featured on the March 1943 cover of *Time* as one of the 50 most important people in Washington.

In his speech accepting the vice-presidential nomination at Lamar, Missouri on August 31, 1944, Truman predicted that

> The end of hostilities may come suddenly. Decisions that will determine our future for years, even generations to come, will have to be made quickly. If they are made quickly and wisely by those who have had years of experience and the fullest opportunity to become well informed with respect to our national and international problems, we can have confidence that the next generation will not have to spill its blood to rectify our mistakes and failures.[122]

Truman pledged to follow the advice of FDR's seasoned advisors. He may (or may not) have departed from FDR's vision a year later, in August 1945, when he consented to the use of the bomb, but there appears to be some disconnect between the actual historical events, including his public statements, and both his alleged reluctance to become President and unsubstantiated claims that he knew nothing about the bomb.

What if Truman did have foreknowledge about the bomb project, as stated in the **Smyth Report**? What might that say about what happened at the Democratic National Convention in July 1944, when the "failed haberdasher from Missouri" was thrust into the role that would make him "accidental president" and undeclared stepfather of the *Little Boy* and the *Fat Man*?

Truman also inherited FDR's difficulties with "Uncle Joe" Stalin. Scholars now reassessing the decisions made by both Truman and FDR with respect to the bomb would do well to consider the subject from the PCnet point of view.

FDR's July journey

On July 6, 1944, the *USS Baltimore* (CA-68) arrived at the Mare Island Navy Shipyard near San Francisco. The *Baltimore* was at Mare Island, a 35-mile drive from Port Chicago, until the evening of July 17, 1944, when she left for San Diego, where she would serve as President Roosevelt's flagship for the cruise to the Honolulu Conference.

The USS *Baltimore* (CA-68), FDR's flagship for the July 1944 cruise to the Honolulu

Conference, was docked at the Mare Island Navy Shipyard, along with the other ships in the COMINCH fleet, from July 6th to July 17, 1944.

The Court of Inquiry that investigated the explosion concluded that there was no evidence of sabotage. It appears that the Court was unaware (or neglected to make note) that the president's flagship had departed from the nearby shipyard, along with the other ships in the COMINCH fleet, within hours of the explosion.

This information is substantiated by unclassified primary documents in the public domain, accessible online; yet these seemingly significant facts do not appear in any history of Port Chicago or of FDR's travels that summer.

Surprisingly, historical accounts of the *Baltimore* sometimes omit the fact that it was FDR's flagship in the summer of 1944. Others gloss over the facts.

- The Wikipedia article, "USS *Baltimore* (CA-68)", describes how she supported the war in the South Pacific, then notes simply that "Returning to the United States in July 1944, she embarked President Franklin D. Roosevelt and his party and steamed to Pearl Harbor."[123]

- Adding the term "Mare Island Navy Shipyard" to the search for USS "*Baltimore (*CA-68*)"* yields first-page hits for 18 October 1944, and even March 1942, but not for July 1944.

- A U.S. Navy document, "*The history of the U.S.S. Baltimore, CA-68*", includes a timeline of the *Baltimore* from April 15, 1943 to March 3, 1946. The entry for June 24, 1944, says "Operating with Carrier Task Group 58.1 made air strike on Iwo Jima. At 2330 Baltimore was detached to proceed at maximum speed to Navy Yard, Mare Island for limited availability.[124]

It would require a careful study to reconcile accounts about various elements of FDR's July journey. In an October 2017 post from the FDR Library blog **Forward with Roosevelt,** Paul M. Sparrow explained that

> After supporting the attack on Iwo Jima, the USS Baltimore steamed at full speed back to Pearl Harbor under a cloak of utmost secrecy. It then traveled to San Francisco for highly secret modifications, and then on to San Diego. Finally on the evening of July 22 there was a surge of activity – and President Roosevelt came on board amidst much pomp and circumstance.[125]

This explanation seems a bit dubious, for several reasons. First of all, since FDR sailed to Hawaii from San Diego, it is not clear why the modifications made to FDR's flagship (which included installation of an elevator and other accommodations) could not have been done at the San Diego Naval Shipyard, a designated U.S. Repair Base.

Secondly, work on the flagship would not seem to have required the presence of the entire COMINCH fleet for three weeks (a fact not reported in the account).

It is also not clear why the *Baltimore* was called from the South Pacific for the urgent, top-secret assignment of serving as the President's flagship instead of outfitting a ship that was already stateside. (The fact that it made two stops may be a clue.)

Sparrow described FDR's trip as

> . . . an historic voyage *shrouded in secrecy* but *intended to generate an enormous amount of press*. FDR was going to Hawaii to meet with Gen. Douglas MacArthur and Admiral Chester Nimitz, the two men responsible for the war in the Pacific. It was also just a few months before the Presidential election, and FDR wanted to show the voters that he was up to the challenge of an unprecedented fourth term in office.[emphasis added]

Sparrow's mention of "much pomp and circumstance" seems to contradict the secrecy and unusual activities described in the first person account by U.S. Navy seaman Douglas MacVane (embedded in the same post) who was serving on the USS Baltimore".[126] MacVane was writing from his limited perspective, of course, and he acknowledged writing from memory, aided by some unspecified records. But he recalled clearly that typical ceremonies were suspended.

In his description of the round trip, MacVane noted that as it headed *toward* the U.S., the ship was unaccompanied and proceeding at an unusual speed; yet it made two brief stops, first at "a small, isolated lagoon" and then at Pearl Harbor.

This raises yet another question. Why did FDR go to Honolulu to meet with Nimitz and MacArthur? They could have traveled aboard the ship that was already in their territory before it headed for the U.S. The USS *Baltimore* was in the South Pacific; MacArthur was somewhere in that vicinity. When the ship stopped at Pearl Harbor, Nimitz boarded briefly and then (apparently) disembarked.

FDR had been ill; he was starting yet another campaign; and the trip would require him to miss the historic Democratic National Convention. So why not stay in the U.S. and send for the Commanders to come to him? Back in the States, MacVane described watching the puzzling goings-on as a variety of laborers remodeled the ship at Mare Island. The crew was unaware, of course, that the work was being done to accommodate the President.

At San Diego, MacVane made note of the impressive group of officers and others accompanying FDR.

> "Never in my young life had I ever seen or witnessed the activities of such an entourage as that which attended President Roosevelt. There were politicians, military officers of every rank imaginable, and various others in attendance. One woman in particular caught my attention. She stood beside the open vehicle and in

contrast to all the rest. But at no time did sight or sound of a dog present itself.[127]

If my memory serves correctly, Baltimore was underway out of Pearl Harbor the evening of the same day we'd arrived.

Secrecy, secrecy, Top Secret, Ultra Secret... Why had we been separated from the rest of the Task Force? Where were we bound? No one was talking.

As I hung around the division gear locker havin' a cup of W.W.II chickory coffee, it was theorized that if we held our present course and speed, we'd run aground on the coast of the United States--somewhere around San Francisco.

For a couple of days (maybe it was three?), Baltimore continued steaming East at a higher that usual speed. And soon, the radar picked up a large land mass, which became larger and larger as time passed.

The "scuttle-butt" (Rumors) at the Division gear lockers had been correct. The good ol' United States of America lay before us. The coast of California was straight ahead. This was a sight to behold--the U.S.S. Baltimore and her crew had been away for a long time.

But, why were we here? Why had we returned state-side when there was a war to fight?

Douglas E. MacVane was a sailor aboard the USS Baltimore, FDR's July 1944 flagship

Back at Pearl Harbor, seeing all the Pacific commanders who stood in formation as the USS *Baltimore* docked, MacVane could not help wondering: who was running the war?

The PCnet explores another possible reason for FDR's sudden, secret trip to the West Coast in July 1944, at the time of the Port Chicago explosion.

As far as journalism and history books are concerned, that fact might have become extinct, but for a tiny clipping in a local newspaper.

Roosevelt Back Day After Blast

SAN FRANCISCO, Aug. 10. — President Roosevelt arrived at the U. S. Marine Base at San Diego, Cal., the day after the disastrous ammunition ship explosion at the Port Chicago Naval Ammunition Depot, 450 airline miles to the north.

A small paragraph in the **Madera Tribune, August 10, 1944** noted FDR's arrival in the San Diego area the day after the Port Chicago explosion. The clip notes the 450 airline miles between the two locations.

The Naval History and Heritage Command explains the route of the *Baltimore*:

> Recalled to the west coast of the United States, *Baltimore* sailed for San Francisco on the latter day, touched at Eniwetok, in the Marshalls, on the 27th, and at Pearl Harbor on 2 July, before she arrived at Mare Island for a limited availability to ready the ship for service as a Presidential flagship. Shifting down the coast, Baltimore arrived at San Diego on 18 July and, on the 21st, embarked President Franklin D. Roosevelt and his party. She then steamed to Hawaii, embarked Admiral Chester W. Nimitz from a tug slowed off Fort Kamehameha on the 26th, and stood proudly into Pearl Harbor with the Presidential colors at the main, while

every ship in the harbor "manned the rail" for this historic visit.[128]

The heavy cruiser USS *Baltimore* (CA-68) arrived at the Mare Island Naval Shipyard on July 6th, 1944, a few days before President Roosevelt finally announced his candidacy for a fourth term. It remained there until July 17th, 1944, departing within hours of the Port Chicago explosion.

The President's special train, the Ferdinand Magellan, had departed Washington, D.C. on the evening of July 13th. FDR spent the night at his home in Hyde Park, New York. He arrived in Chicago on the 15th. In a brief meeting with Democratic Party leaders, he finally answered the question America had been asking: Truman was his candidate to become the Vice-Presidential nominee in the Democratic National Convention when it convened that week.

The President's train made several brief stops, including one at Tucumcari, New Mexico at around 5 P.M. on the 17th, before it appeared in San Diego in the wee hours of the 19th.[129]

DECLASSIFIED
Authority: E.O. 13526
By: NDC NARA Date: Dec 31, 2012

18 July, 1944. (Zone - Plus 7 1/2)
Underway in company with Task Group 12.1, enroute from Mare Island, California to San Diego, California. OTC is CTG 12.1 in U.S.S. BALTIMORE (CA68). 1334 Passed Point Loma abeam to port, distance two miles. 1452 Moored in berth #18, San Diego, California. Ships present; various units of U.S. Pacific Fleet plus various yard and district craft. SOPA is Commander Task Group 12.1, Commanding Officer U.S.S. BALTIMORE (CA68).

Positions:	0800	1200	2000
Lat.	32-47-30 N	32-35-30 N	
Long.	119-08-30 W	117-45-45 W	

19 - 20 July, 1944. (Zone - Plus 7 1/2)
Moored in berth #18, San Diego, California.

The USS *Baltimore* (CA-68) and other ships in the COMINCH fleet arrived in San Diego, from Mare Island, between July 19th and 20th, around the same time of FDR 's unannounced arrival.

The President's itinerary was not published during this secret journey, as it had been on his 1942 trip to the West Coast. According to one newspaper account (which seems to corroborate MacVane's recollection), there was none of the usual fanfare when he arrived in San Diego.

Of course, the Secret Service knew the President's whereabouts, which could explain why the COI ruled out sabotage as a cause of the Port Chicago explosion, and why there appeared to have been no concern that his flagship may have been in harm's way although it had just left the vicinity. (This contrasts with an event in 1942, when FDR was departing from a defense factory. There had been an accident on the railroad tracks ahead, leading the Secret Service to delay his journey until they could confirm that the accident was not an attempt on FDR's life.)

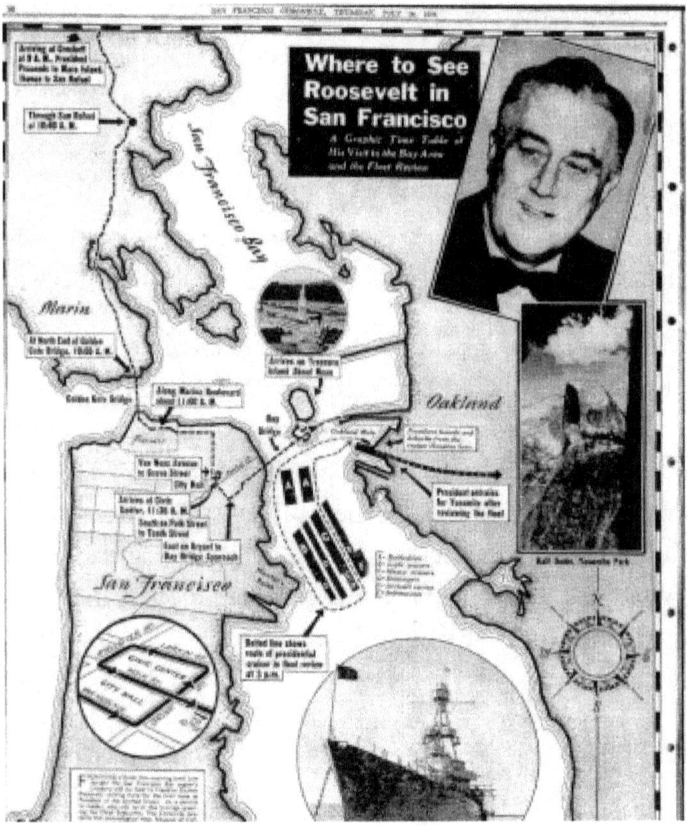

The itinerary for FDR's 1938 West Coast trip was published in newspapers. The President also toured the Bay area in 1942.

The President's flagship

Some histories of the USS *Baltimore* make no mention of July 1944 or the fact that the *Baltimore* was President Roosevelt's flagship for the trip to the Honolulu Conference. They do report, though, that the ship was at Mare Island in October 1944. The omission might seem insignificant in other contexts, but is of interest to the PCnet because

1. Transporting the President would seem to be one of the key events in the ship's history. Indeed, MacVane describes it as "The one event that topped the very extensive and impressive combat service of U.S.S. *Baltimore* (C.A. 68)." The fact that the very historic July 1944 trip was omitted from some accounts may signal a secret explanation of why the *Baltimore* went to San Francisco and not directly to San Diego.

1. The omission is one of the many important gaps in early nuclear history. One of the ways the narrative-makers intentionally skewed the history of the bomb was by omitting key facts and details.
2. FDR was on the West Coast in July 1944, at the time of the Port Chicago explosion. His whereabouts, and that of his ship, are of core importance to the PCnet. From a traditional point of view, the discrepancy may be overlooked or deemed a benign error; but in an investigation of the PCnet, no stone can be left unturned. The results would be skewed either by *assuming* the account is accurate and complete, or by dismissing the omission as of minor concern. (This illustrates the importance of using a variety of sources, especially official records and primary documents.)
3. A commemorative calendar detailing the ship's history designates San Diego as the destiny. From the PCnet point of view, the apparent discrepancy is another red flag. Small details like this contribute to the questions that drive historical research. Reading the various reports, a historian might ask, for example

- *"Did the ship sail to San Francisco or did it go directly to San Diego?*

- *How might this be an important detail?*

- *What other details of this account are worthy of further inquiry?"*

Such questions are important to the PCnet. So is the omission of small but significant details. There may be nothing to hide in the simple history of the ship; but to cover up a nuclear explosion, there would be a great deal of information to hide.

1. There are also discrepancies among various sources about whether the *Baltimore* arrived in San Diego on the 18th or the 19th. The significance of this detail becomes obvious by comparing the fanfare at Honolulu with the surreptitious arrival of the ship at San Diego, where basic protocols were ignored, including the standard announcement that the President was boarding the ship.

With a less secretive President, the assumption that this was a simple security measure might be reasonable, but FDR loved intrigue. On at least one occasion, he appeared to be in two places at one time.[130]

FDR's itinerary

On July 13th, two days after he finally announced his candidacy for a fourth term, President Roosevelt's train, the *Ferdinand Magellan*, left Washington, D.C., arriving at his home in Hyde Park, New York, where he spent the night on the 14th. From there he embarked on a sudden, secret, cross-country trip to the West. Coast. He stopped in Chicago, Illinois on the 15th to meet with Party bosses preparing for the Democratic National Convention, set to begin that week.

Take note here; language is important. A simple, well-crafted statement can hide secrets in plain sight. The first sentence *here*, for example, could be a ruse: it only *implies* that the President was aboard the train when it left Washington. He may or may not have been.

FDR's slippery secretiveness, and his sudden secret trip to the West Coast during the week of the Port Chicago explosion calls for a careful study of his itinerary from July 14th to July 21st, 1944. Details matter, especially in an investigation of the PCnet.

FDR liked to see things for himself. He didn't mind taking a risk to do so. For example, in November 1943, he was on his way to the Tehran conference, aboard the *USS Iowa*, which was escorted by the USS *William D. Porter*. The "Willie Dee", perhaps the most ill-fated ship in history, had already had several mishaps when it accidentally fired a live torpedo at the *Iowa*. Others took cover, but President Roosevelt asked his Secret Service agents to wheel him closer to the railing so he could observe the action.[131]

In February 1939, shortly after Bohr announced the discovery of nuclear fission, FDR slipped off to the Caribbean to observe Fleet Problem XX from the battleship USS *Pennsylvania*. The Fleet Problems were mock battles that the U.S. Navy used as part of their training exercises between 1923 and 1940.[132]

The atomic bomb would replay the old inter-service rivalry, exemplified by the Billy Mitchell scandal of the 1920s when the question was how air warfare would impact the Navy. In the spring of 1939, there was yet another new technology in the offing.

FDR's ongoing health problems

President Roosevelt looked ghastly in the photograph taken on the night of his acceptance speech. People were surprised at his appearance. It wasn't just that the picture was unflattering; the President looked pallid. But why?

FDR's appearance in the photo taken during his acceptance speech on the night of July 20, 1944 raised fresh concerns about his health. Of course, there were no public reports about the sudden attack of pain he had suffered earlier that day. In a letter to First Lady Eleanor Roosevelt, he changed the timing and dismissed the incident as a case of the "collywobbles."

At the beginning of the year, President Roosevelt had been so ill that he had spent the month of April away from Washington, recuperating at Hobcaw

Plantation, the South Carolina home of his good friend, financier Bernard Baruch.[133]

Rumors about his health persisted. Public information releases were misleading. FDR maintained his habitual secrecy; the press maintained the ongoing "gentleman's agreement" that had long kept the public unaware of the extent of his disability. Today, historians and biographers continue to publish new updates on his multiple diagnoses.

On July 13th, when he left Washington, D.C. to spend the night at Hyde Park, FDR had been back in the White House for two months, his work schedule drastically reduced.

On July 10, 1944, Frank Lahey, one of the doctors on the team that had attended him at the Bethesda Naval Hospital back in March, recorded his belief that FDR would not likely survive another four-year term.[134] (The document, which Lahey wrote to cover his actions should they ever come into question, remained secret for decades.)

Historians, journalists and political opponents have speculated about the "real" reasons for FDR's sudden secret trip to the West Coast. Some said it was a campaign stunt, that he was out to prove that he was fit for another term. Others believed he took the trip as an opportunity to rest and relax. Officially, he was headed to San Diego, where he would board the U.S.S. *Baltimore* (CA-68), the heavy cruiser that was his flagship for the trip to the Honolulu Conference, where he met with General Douglas MacArthur and Admiral Chester Nimitz, his commanders in the South Pacific.

First Lady Eleanor Roosevelt traveled with her husband for part of the journey. He was also accompanied by his cardiologist, Dr. Howard Bruenn, and his regular doctor, Admiral Ross McIntire. Bruenn, who had been brought on board back in the spring, insisted that FDR cut back his hours and take it easy. He wanted to inform the President that his heart was weak, but McIntire was concerned that the bad news might have a detrimental impact on the President's health.

Though concerns and rumors about his health persisted at the time, the fact that he was accompanied by his doctors *suggests* that they at least acquiesced to the leisurely cross-country railroad trip and cruise.

When the July journey became known to the public, Admiral Wilson Brown, FDR's Naval aide, explained that the leisurely cross-country trip had indeed given the President an opportunity to rest and work on his upcoming speech.

Brown kept written records of FDR's itinerary. The PCnet takes note that there are no entries between 5 p.m. on July 17th and his arrival in San Diego – with or without fanfare, depending on which sources are deemed most credible – in the wee hours of July 19th.

On the morning of July 20th, FDR observed a Marine landing operation at Camp Pendleton. That was apparently after he suffered a sudden bout of pain that left him writhing on the floor of his train.[135]

That evening, he gave his acceptance speech from his train, announcing to the public that he was broadcasting "from a Naval base on the West Coast." On the 21st, he embarked for the cruise to Hawaii, arriving on the 26th.

Official U. S. Navy photograph

AT HAWAII in 1944 President talked with General of Army MacArthur, Fleet Admirals Nimitz and Leahy.

Pictures from the Honolulu Conference may *conceal* a thousand words. The FDR they reveal is not the same man eyewitnesses reported seeing.

One wounded veteran described his surprise when the President was wheeled into the hospital ward in Honolulu.

> ". . . I kept watching the corridor, and suddenly the mob parted slightly, and I saw the President in his wheelchair. His chin drooped on his chest, and his face sagged in a mass of baggy wrinkles. He looked ghastly, almost unrecognizable as the exuberant, vibrant President I had often seen in news photos and on the newsreels.

A moment later the commanding officer of the hospital took one step into our room, announced formally, "Gentlemen, the President of the United States," and then stepped back out again. The President was wheeled into our room. I was stunned. You would have thought he was being wheeled onstage at Soldier Field, Chicago, before fifty thousand people. His head was thrown back, that famous broad, open Roosevelt smile on his face had wiped out the sagging wrinkles, and his right hand was raised high in cheery greeting. A complete transformation, and all for the benefit of two wounded Marines. He stayed for a few minutes and chatted, but I no longer remember what he said except that at one point he told a dull joke at which he laughed delightedly, and then he was gone. A complete transformation, and all for the benefit of two wounded Marines. . ."[136]

According to John T. Flynn, author of **The Roosevelt Myth**, the generals were shocked at the President's ghastly appearance, especially after his five-day cruise.[137]

Why the sudden trip?

Historians speculate about the "real" reasons for FDR's sudden, secret trip in July 1944. Officially, he was on his way to the Honolulu Conference, where his goal was to get his South Pacific commanders to devise a united plan for the war; and possibly, some say, to confront Douglass MacArthur, a potential Republican rival in the upcoming election. Perhaps it was part of a campaign strategy, to prove that he was fit for a fourth term. Of course, the complex President may have had multiple objectives.

And it is equally likely that he had a hidden motive, one so well camouflaged that, after almost 80 years, it has yet to be revealed. Camouflage is clever. Things that are camouflaged can be hidden in plain sight.

- Like the USS *Baltimore* sitting for weeks at the Mare Island Shipyard, departing within hours of the Port Chicago explosion.

- Like FDR taking a trip to the West Coast, making a phone call from Tucumcari on the 17th and appearing next (without fanfare) in San Diego on the 19th.

- Like getting the "collywobbles," for no apparent reason; and having that incident go unreported, as the dramatic decline in his health continued right up until his death.

None of these facts is proof that FDR made the trip to the West Coast so that he could be on hand to observe the clandestine test of his new exponential secret weapon. By the same token, this scenario accounts for these little-known but significant details, and for related secrets about other people, like Albert Einstein, whose role in the creation of the bomb merits further review.

Effects, and causes

One way to analyze a historical event is to study its effects and try to determine the extent to which they might have been accomplished by design. In the PCnet, we have at least three historic events for which the most obvious and lasting results have been traditionally overlooked.

The bomb was not used against Hitler either in combat or as a deterrent to his use of an atomic bomb. It can be argued, on the basis of evidence, that that was never its purpose. In both his April 23, 1945 letter to the Secretary of War and his memoir, **Now It Can Be Told,** General Leslie R. Groves stated that "The target is and was always expected to be Japan." Japan was also identified as the target in the May *1943* meeting of the Manhattan Project Target Committee.

Whether or not the two atomic bomb attacks really ended the war in Japan is still up for debate. The immediate effect of the attacks on Japan was that use of the bomb established the U.S. as the world's first atomic power (and, of course, the only nuclear power at that time).

Historians disagree about the extent to which Truman's decision to use the bomb was actually a means of managing Joseph Stalin. But to appreciate the rationale for *creating* the bomb, it is necessary to consider what *FDR* thought about the inherent advantage of being the first nation to obtain the superlative weapon that changed the world. Although he did not live to see his vision materialize, writers have proclaimed for decades that the post-war world order was designed by Roosevelt.

The Port Chicago explosion may have been one of the most

consequential of all the little-known events in modern history. Prior to the explosion, the scientists and engineers working on the Manhattan Project lacked the data they needed to design, produce and test the weapon.

July 17, 1944 marked a turning point in Manhattan Project history, thanks in part to the critical information obtained from the effects of the explosion, and also because the first atomic bomb model, the *Thin Man*, was shelved on that date, leading to the restructuring of the Project and a shift in production priorities.

It will require further research to determine to what extent and in how many ways nuclear history was impacted by the Port Chicago explosion.

The Port Chicago mutiny trial, largest mass court martial in U.S. Naval history, was completely within the control of the Navy. As we have seen, the situation could have been handled in other ways. The mass trial upstaged one of the largest and most consequential explosions in history. The court martial soon became the defining element of the Port Chicago disaster.

Every July, a small group of people gather at the Port Chicago National Memorial to "Remember Port Chicago." That phrase applies in part to those who died in the blast, but it was also the slogan for the NAACP campaign to exonerate the sailors who were convicted of mutiny. The legacy of that campaign is reflected in the new legislation on their behalf the Congress considers each year.

As historians and scholars reconsider the history and the future of the bomb, they would do well to include a comprehensive study of the Port Chicago explosion.

Einstein's torpedo

Einstein was one of several influencers who believed that the only way to achieve world peace was through a one-world government. It is likely that this is one of the topics Einstein discussed with Churchill on his visit to England in January 1933. It is a little-known fact that Einstein also visited with FDR. In January 1934, he and his wife were overnight visitors at the White House. FDR and Einstein both loved sailing. According to Einstein biographer Banesh Hoffman, "they spoke, too, of the growing darkness in Europe."[138]

After the war, Einstein explained that his contribution to the bomb project "consisted in a single act: I signed a letter to President Roosevelt." (Signed, not wrote.)

Given the high security of the project and the policy of "compartmentalization", which limited the flow of information on a 'need-to-know' basis,' it is possible that Einstein — like most people who worked for the Project — was kept in the dark about other ways his work contributed to the development and research of the bomb. For example, the coincidental timing, continuing secrecy, and documentary evidence all *suggest* that Einstein's work on torpedo design as a consultant for the Navy may have been related to Parsons' quest for a nuclear torpedo.

Torpedoes were still a new technology. Insufficiently tested, many of them failed during the first years of the war.[139] One biographer speculates that Einstein's work on torpedo design was in connection with the "great torpedo scandal." The Navy did consult leading experts, including Einstein, with regard to the problem. His contract began in May 1943—notably the same month Bush recommended Parsons to head Manhattan Project ordnance.

Einstein's contribution to the torpedo scandal was of no particular value; however, that is an unlikely reason why his work on torpedo design—which may seem contradictory to his reputation as a pacifist—is not well-known. It is helpful to consider the timing of his work in the context of other contemporary research and in light of the people with whom he had contact.

For example, details about the first post-war nuclear tests, *Operation Crossroads*, reveals a significant connection that is an unlikely coincidence. Parsons was Technical Director of what he called "the largest laboratory experiment in history." Commodore Stephen Brunauer served as Officer-in-Charge under Parsons' direction. As a lieutenant during the war, Brunauer had been the Navy's primary liaison with Einstein.

It is clear from several sources that Einstein did know about the Manhattan Project. For one thing, some of his closest colleagues were involved in the theoretical research. For example, John von Neumann, the eminent mathematician, was his colleague at the Institute for Advanced Studies at Princeton. Einstein consulted von Neumann on at least one occasion in the summer of 1943.

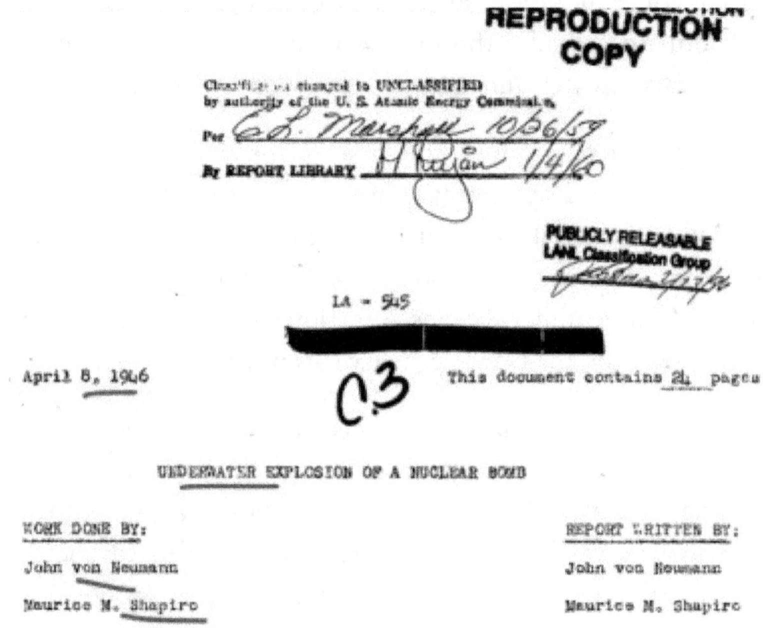

REPRODUCTION COPY

Classification changed to UNCLASSIFIED
by authority of the U. S. Atomic Energy Commission

Per ____ 10/26/57

By REPORT LIBRARY ____ 1/4/60

PUBLICLY RELEASABLE
LANL Classification Group

LA - 945

[redacted]

April 8, 1946

C3

This document contains 24 pages

UNDERWATER EXPLOSION OF A NUCLEAR BOMB

WORK DONE BY:

John von Neumann

Maurice M. Shapiro

REPORT WRITTEN BY:

John von Neumann

Maurice M. Shapiro

John von Neumann and Maurice Shapiro authored an April 1946 study on underwater explosion of a nuclear bomb.

Parsons relied on von Neumann's studies of the shock wave of an underwater blast. When von Neumann visited Los Alamos that September, he and Parsons spent two days in an intense consultation. The official explanation is that they were working on implosion, a new idea about how to detonate a nuclear weapon. However, implosion did not become a priority for the Project until July 1944, after the *Thin Man* bomb was shelved.

In the fall of 1943, Parsons was still pursuing underwater delivery. Further research may show that von Neumann also contributed to the work on an atomic torpedo. Parsons also consulted von Neumann in 1946 with regard to *Operation Crossroads*.

John von Neumann and Maurice Shapiro co-authored an April 1946 study on underwater explosion of a nuclear bomb. Both men were involved with Parsons' research during the Manhattan Project; Shapiro wrote at least one of the reports on the Port Chicago explosion.

An intricate network

Few sources go behind the surface of the bomb story to study the connections between the men involved. It can be confusing. It is also informative.

- In March 1939, Enrico Fermi, Leo Szilard and other scientists hoping to gain support for uranium research met with officials from the Navy, including Ross Gunn, Technical Director of the Naval Research Laboratory (NRL) from 1939 to 1946. George **Pegram**, a dean at Columbia University, arranged the meeting. Pegram contacted Admiral S.C. **Hooper**. Hooper called the **Navy Department**. Charles **Edison**, son of the famous inventor, took the call. Gunn subsequently wrote a letter to Szilard saying the Navy was interested in the scientists' work but could not support their uranium research. Instead, the Navy Research pursued the use of uranium to power submarines.[140]

- President **Roosevelt** had appointed Charles Edison as Assistant Secretary of the Navy on January 18, 1937; he became Acting Secretary on January 2, 1940. Edison was the son of **Thomas Alva Edison**, the famous inventor. During World War I, when FDR was Assistant Secretary of the Navy, the senior Edison had pushed for better military technology, offering, for example, to develop an airplane that could drop a bomb with precision.

- In July 1915, Navy Secretary Joseph Daniels, FDR's boss, invited Edison to chair a new Naval Board that would promote military science and technology. The Board was administered by FDR. In July 1923, the Navy, following a recommendation from Edison, established the **Naval Research Lab**.

- William S. **Parsons** served in the **Bureau of Ordnance**, Navy Department, Washington, DC, from July 1933 until May 1934, acting as liaison officer for the Bureau at the **Naval Research Laboratory**, Anacostia, DC.

• In 1940, a young physicist named Philip **Abelson** went to work at the **Carnegie Institute of Washington** (headed by **Vannevar Bush**). Abelson worked for **Merle Tuve**. While there, Abelson began conducting independent uranium research. In September 1940, Tuve went to the **Dahlgren Naval Proving Ground** in Virginia, where he worked on development of the proximity fuze with **Lieutenant Commander William S. Parsons**. In June 1941, Abelson went to work at the Naval Research Lab with **Ross Gunn**.

• In April 1942, **Parsons** became Special Assistant to the Director of the OSRD – **Bush**'s right-hand man; now **Tuve** worked for Parsons.

• Tuve was an original member of the Uranium Committee set up by FDR October, which first met on October 21st, 1939 — a little over two weeks after Sachs made his presentation to FDR. Gunn, another member of the UC, had worked with Parsons and Abelson at the Naval Research Lab.

The traditional narrative says that because the Navy's sole interest was in a nuclear-powered submarine, they turned down the scientists' request for uranium research support. The narrative also says the Navy was deliberately left out of the Manhattan Project. But transporting the bomb to the Pacific was perhaps the least of the Navy's contributions to the bomb project.

(Left to right) Ross Gunn, William S. Parsons, Philip Abelson worked together at the Naval Research Lab before the discovery of nuclear fission. Gunn was on the original Uranium Committee established by FDR in October 1939, when Parsons was Special Assistant to Bush.

Many naval personnel worked for the Project, including some key officers recruited by Parsons himself. Meanwhile, according to Tuve, the Navy was funding research at CIW. Further research may reveal the extent to which this research supported the Project; but the interconnections among significant people might warrant additional study.[141]

Einstein, Szilard, Sachs; Abelson, Gunn, Parsons; Bush, FDR . . . one contact leads to another; next thing you know, you have a network.

One idea – especially one shared by and among a small, like-minded and resourceful group of innovative people – leads to another. One way to study the secret Manhattan Project might be to trace the development of the ideas that made the bomb possible. Another way is to examine the ideas associated with the theory that a nuclear weapon may have been tested at the Port Chicago Naval Ammunition Depot on July 17, 1944.

America's bomb

The idea of an atomic bomb was of a weapon that was like none other. It was an idea that changed the world. The Manhattan Project is still heralded as a superlative achievement, a model government program. The scope of the Project, the technological achievement, the coordination of great minds, the use of expansive resources and the military prowess are all iconic of American greatness, American supremacy.

What is seldom acknowledged, though, is that the *secret* government / academic / military enterprise that changed the world also exemplifies the definition of conspiracy. It was, by definition, a secret plan by a small but powerful group who collaborated to achieve a dubious goal by any means necessary.

These two statements of fact are equivalent; the thought is sobering, especially in context of Parsons' acknowledgement, in his April 1950 speech to the Naval War College, that during the war, the United States had operated under a robust dictatorship. He attributed U.S. military success to the fact the military had enjoyed "absolute priority of effort" (which included manpower and material resources).

> We operated under a *semi-dictatorship* in World War Two. **We had absolute priority of effort, and we got results like the Manhattan District**, production of aircraft and production of fleets. We then demobilized and concentrated on automobiles, television and like things. But we forgot that Russia had not demobilized and was still operating under **a *dictatorship* more rigid and perhaps as dynamic as the one that we had operated under in World War Two.**[142] [emphasis added]

Daniel Ellsberg, author of **The Doomsday Machine: Confessions of a Nuclear War Planner**, was 13, a ninth grader at a private school in Michigan in 1944 when he first heard of nuclear fission[143]. His social studies teacher described the weapon's potential and gave the students an assignment. They were to write an essay on how an atomic weapon might impact the world in the future. Even then, Ellsberg says, "everyone in the class had arrived at much the same judgment. It seemed pretty obvious: the existence of such a bomb would be bad news for humanity."

Ellsberg, the Department of Defense whistleblower who leaked the Pentagon Papers in 1971, notes the irony that to Americans hearing about the bomb for the first time on August 6, 1945, after it wiped out Hiroshima, the weapon "was 'our' instrument of American democracy." They got this idea from the postwar publicity campaign, the spin that the bomb was a gift from God.

Ellsberg says,

> Whether rightly or wrongly, we are the only country in the world that believes it won a war by bombing—specifically by bombing cities with weapons of mass destruction, firebombs, and atomic bombs—and believes that it was fully justified in doing so. It is a dangerous state of mind.[144]

Before Americans bought the *story* of the bomb, they literally and unwittingly bought the bomb itself. Though paid for with American taxes and produced with American labor, the weapon was a complete secret to Americans during the war.

Secrecy remains the hallmark of nuclear policy. There has been no public referendum on the bomb. Nuclear historians express concern that although the threat of nuclear annihilation is great and grows continually greater, American citizens today remain unaware of and uninvolved in developments in nuclear policy.

That being the case, it is reasonable to believe that a nuclear explosion at Port Chicago could have been conducted in secret during the war, and covered up for almost a hundred years.

A small group of men thought it was necessary to build the bomb. It was necessary to use it. And, although it is not reflected in the traditional narrative – which was deliberately deceptive – a variety of compelling evidence suggests that Manhattan Project engineers may have found it necessary to conduct a secret test of the weapon.

This cautionary tale, with its potentially apocalyptic ending, goes unheeded. By contrast, in an opinion piece in the *New York Times* columnist David Leonhardt hailed "The FDR Approach". He admired the courage and boldness with which FDR took risks during the national emergency.

> Franklin D. Roosevelt repeatedly broke with tradition — and endured confrontation with the courts — to fight the Great Depression. His administration, working closely with business, also threw out the normal bureaucratic procedures to build World War II ships, planes, tanks, bombs and other matériel with stunning speed.[145]

To Leonhardt, FDR responded aptly to the wartime emergency. This contrasts, however, with the phrase accompanying Truman's picture on the March 1943 cover of *Time Magazine*, which insists that even and possibly especially in wartime, "a democracy has to keep an eye on itself."

As a wartime President with the singular opportunity to build an exponential, world-changing weapon, FDR was bound neither by precedent nor by his own moral appeal to other nations, at the outset of the war, not to bomb civilians. Nor was he subject to the scrutiny of the press and the public. This lends substance to Parsons' observation that the U.S. operated under a dictatorship during the war.

In his first inauguration speech in 1933, FDR asserted his "firm belief that the only thing we have to fear is fear itself – nameless, unreasoning, unjustified terror which paralyzes needed efforts to convert retreat into advance." Still, some of his supporters, including First Lady Eleanor Roosevelt, were wary of the meaning behind some of his other statements about the flexibility of the Constitution and Democracy.

> Our Constitution is so simple and practical *that it is possible always to meet extraordinary needs by changes in emphasis and arrangement without loss of essential form.* That is why our constitutional system has proved itself the most superbly enduring political mechanism the modern world has produced. It has met every stress of vast expansion of territory, of foreign wars, of bitter internal strife, of world relations.
>
> It is to be hoped that the normal balance of executive and legislative authority may be wholly adequate to meet the unprecedented task before us. But it may be that an unprecedented demand and need for undelayed action may call for temporary departure from that normal balance of public procedure. I am prepared under my constitutional duty to recommend the measures that a stricken nation in the midst of a stricken world may require. These measures, or such other measures as the Congress may build out of its experience and wisdom, I shall seek, within my constitutional authority, to bring to speedy adoption.
>
> But *in the event that the Congress shall fail to take one of these two courses,* and in the event that the national emergency is still critical, I shall not evade the clear course of duty that will then confront me. *I shall ask the Congress for* the one remaining instrument to meet the crisis—*broad Executive power* to wage a war against the emergency, as great as *the power that would be given to me if we were in fact invaded by a foreign foe.*
>
> For the trust reposed in me I will return the courage and the devotion that befit the time. I can do no less.[146] [emphasis added]

By 1942, the practices that had become standard operating procedure for FDR allowed him to secretly authorize the Manhattan Project. There was, after all, a *total war* on. Bypassing democratic "bureaucracy", it seemed, was still the necessary thing to do.

FDR's broadened executive power would later be reflected in the first recommendations for Presidential authority over nuclear weapons. As noted earlier, in their June 1947 report on *Operation Crossroads*, "The Evaluation of the Atomic Bomb as a Military Weapon", the Joint Chiefs of Staff Evaluation Board proposed an amendment to U.S. first-strike policy, with the corresponding recommendation that should another nation *contemplate* an attack on the U.S., the **President** should be empowered to make the instant, **unilateral decision** to make an offensive nuclear attack.

> (6) *National defense requirements of the future are only those of the past*; any aggressor must be overcome with superior force. But, *where in the past, the duty of the President, as the Commander in Chief has been restricted* (before formal declaration of war) to action only after the loss of American lives and treasure, **it must be made his duty in the future to defend the country against *imminent or incipient* atomic weapon attack.**[147] [emphasis added]

Leonhardt, who was commenting on the government's response to the COVID pandemic, suggested that President Biden should follow FDR's example.[148] In today's climate of succeeding crises, his argument is all too appealing – that is, without the context of the evidence revealed by the PCnet.

Many people assume that the precarious unilateral control of the bomb by the President of the United States evolved as a result of post-war political maneuvering, but the evidence shows how that the precedent was established under FDR. The bomb that was a world-changing weapon also had permanent impact on U.S. military policy and on the form and practice of American democracy.

Reviewing Einstein's well-kept secrets

The history of the atomic bomb is flawed. The narrative was designed to sell the bomb to the people. One way to do this was through deception and omission of information of both obvious and unapparent importance.

For example, with all the volumes of books on FDR, on Einstein and on the bomb, why is there so little coverage of the historic occasion when two of the most influential men of the 20th century had a sleepover at the White House?

When reconsidering the likelihood that Einstein may have contributed to the bomb project, in ways yet to be revealed, there are several little-known facts to bear in mind:

1. Einstein did not write the so-called Einstein letter. He later said his only act in support of the bomb was to sign the letter; but he went along with the post-war ruse that led the public to believe the bomb project resulted from *his* letter warning FDR of a possible German bomb.

1. Einstein and his wife spent the night at the White House with the Roosevelt's in January 1934. Given their positions of leadership, their shared values with respect to world government, and the state of world affairs at the time, there is every reason to wonder how the private and unreported conversation between the two powerful influencers may have impacted world history, and possibly contributed to the development of the bomb.

1. Einstein, who declared himself a *pragmatic* pacifist, gladly served as a consultant for the U.S. Navy during World War II. This was, after all, the refugee scientist's adopted homeland.

1. Einstein was hired in May 1943; he was recruited by Lieutenant Stephen Brunauer, who was his liaison with the Navy Bureau of

Ordnance (BuOrd). Parsons joined the Project that same month. In the past, Parsons had served as liaison between BuOrd and Vannevar Bush (OSRD). During the war, he was liaison between BuOrd and the Manhattan Project. In 1946, Brunauer served as Officer-in-Charge of *Operation Crossroads*, over which Parsons served as Technical Director.

1. Einstein's primary assignment for the Navy was work on the design of torpedoes. During that same period, Parsons was pursuing the possibility of underwater delivery – i.e., an atomic torpedo.

1. Einstein was an inventor; he and his fellow physicist, Leo Szilard, held a patent on a wireless refrigerator. Szilard, the actual author of the famous "Einstein" letter to FDR, collaborated with Alexander Sachs, whose own letter was included in the dossier he presented to FDR on October 11th and 12th, 1939.

Further complicating the myth of the Einstein letter is the unsubstantiated, unrealistic false claim that Einstein was barred from the project.

There are several reasons to challenge the claim. First of all, there are too many different stories about it, and none is supported by adequate documentation. Some sources say Einstein was barred from the Project because of his extensive FBI file, yet no specific concern about him seems to have warranted barring the brilliant, world-famous physicist from the urgent Project. By contrast, General Groves intervened to get security clearance to allow J. Robert Oppenheimer, a known Communist sympathizer, to head the Los Alamos laboratory.

Some sources say Bush thought Einstein would be too indiscreet about the Project; yet to this day, few people know that the famous scientist was a consultant for the Navy during the war, that he worked on torpedo design, or that he once spent the night at the White House.

Einstein worked on torpedo design for the Navy at the same time that Parsons was investigating the possibility of underwater delivery. If the ordnance leader of the Manhattan Project was considering the possibility of a nuclear torpedo (or some similar, unheard-of weapon) it would seem self-defeating to ban one of the leading theoretical physicists, especially on such flimsy excuses.

Historians searching government archives are apt to miss tiny clues from lesser-known sources. Two examples (cited earlier) are a soldier who described FDR's appearance at his hospital ward in Hawaii in July 1944 and a sailor (MacVane) who reported his experience on the ship that took FDR to Hawaii.

In a similar obscure story, Charles Lee Sr., of Seadrift, Texas, a former member of the U.S. Army Air Force, reported that Einstein was one of the scientists he had chauffeured at Los Alamos. Melony Overton interviewed Lee for the May 30, 2017 issue of a local news outlet, **The Port Lavaca Wave.**

> "He [Einstein] would often step into the car with his shoes on the wrong feet with one sock on and one sock off," Lee said. "He was a nice fellow, but strange. He would often talk theory with us, but of course we couldn't understand what he was talking about. I often wondered how we won the war with him on our side."
>
> "He had a thick, German accent. He was brilliant, though. You know they say there's a fine line between brilliance and insanity. I think it was true. He seemed to live within his own head most of the time. He would often answer a question, or not answer it. He would walk off and leave in the middle of a conversation."[149]

The myth that the *Navy* was kept out of the project may be related to the alleged ban against Einstein's participation. That decision, too, is sometimes ascribed to Bush, who is said to have resented the way he was treated by top Naval officials. But Bush said it was FDR's decision.

Bush did not explain why "the Sailor in the White House" would have taken that stance, especially since FDR (a Navy man) knew about the legendary rivalry between the Army and the Navy and also understood how military science and technology could determine the nature of warfare.

The PCnet provides one good reason for the Navy to be barred from *reports* about the Project: investigation of the Navy leads to Parsons, and Parsons to the PCnet.

Einstein himself perpetuated the deceptive Einstein letter story. After the war, he posed for a picture "reenacting" the signing of the letter. It could not have escaped him that the purpose of the picture was to sell the bomb as the product of a beloved genius who was a well-known pacifist and humanitarian. By contrast, few sources show the real-time photo of Einstein posing with some of his Navy handlers in his office at Princeton's Institute for Advanced Study on July 24th, 1943.

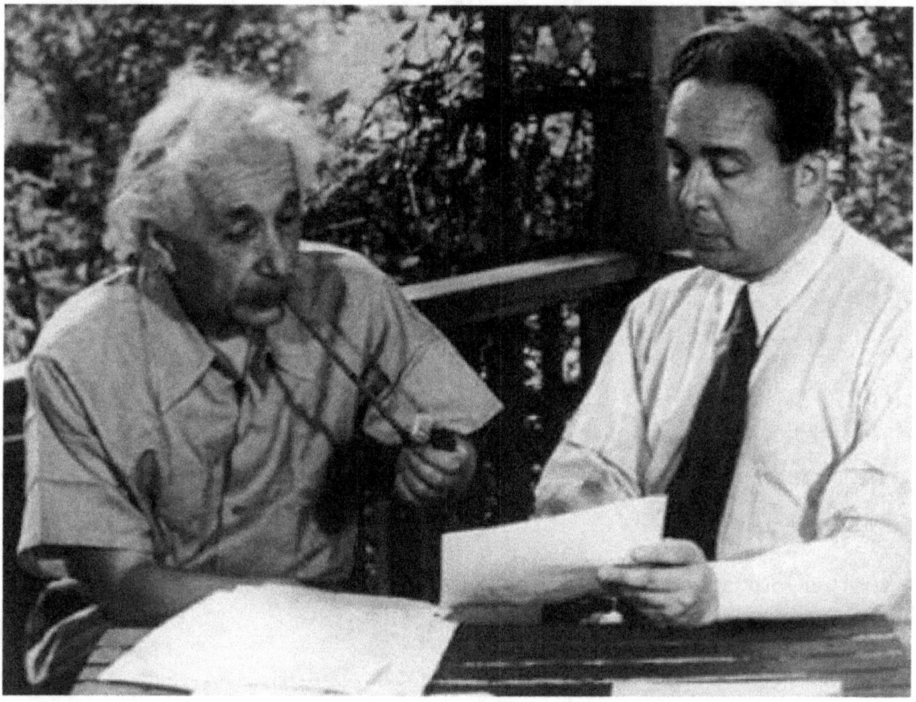

Einstein and Szilard star in a publicity photo designed to sell the "Einstein letter" story. Further research may explain why they posed for the photos knowing they supported the misleading narrative.

Photo # 80-G-42919 Dr. Albert Einstein in his study at Princeton, N.J., 24 July 1943

Einstein with Navy officers at his Princeton office, July 1943. Unlike the 'reenactment' of the Einstein letter photo, this contemporaneous picture has not been widely publicized.

If the Navy was left out of the project, why did a Navy captain serve as bombardier (or weaponeer) aboard the airplane that dropped the bomb on Hiroshima? Why, in fact, did a Navy captain lead the ordnance division of the Manhattan Project?

Another look at the content of the "Einstein letter" addresses that question.

“A single bomb, carried by boat…”

The new evidence revealed through research of the PCnet updates the flawed narrative concerning Einstein and the bomb. These little-known facts make it less credible that the world’s most famous scientist would have been barred from participation in the urgent and intensive effort to produce the world’s first atomic bomb, a weapon of unprecedented power that “harnessed the power of the sun.”

Einstein was a pragmatic pacifist, happy to do what he could to support his adopted country. He was hired as a consultant to the Navy in May 1943. That same month, Parsons was recruited to head the Ordnance Division of the Manhattan Project. Einstein may not have known the precise purpose of his work, but there are limited uses for torpedoes.

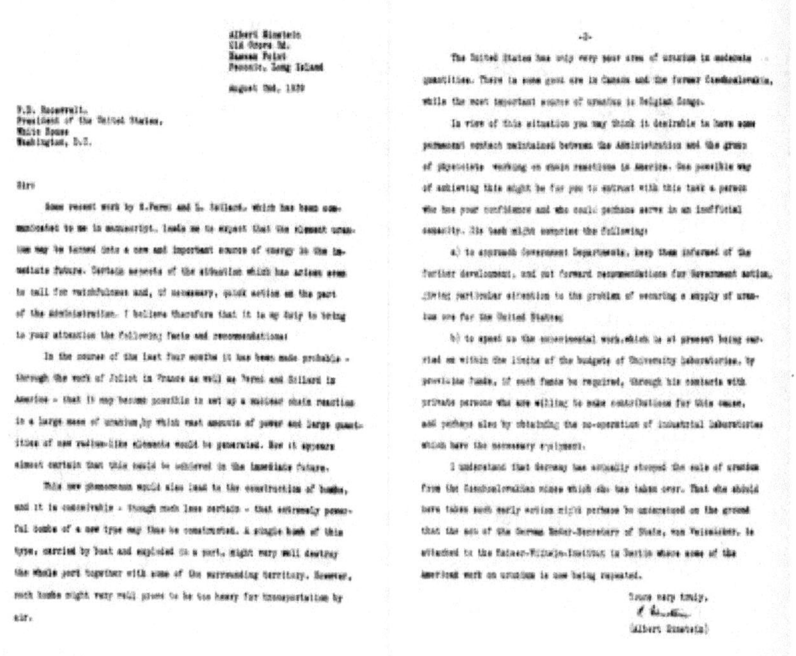

Albert Einstein
Old Grove Rd.
Nassau Point
Peconic, Long Island

August 2nd, 1939

F.D. Roosevelt,
President of the United States,
White House
Washington, D.C.

Sir:

Some recent work by E.Fermi and L. Szilard, which has been communicated to me in manuscript, leads me to expect that the element uranium may be turned into a new and important source of energy in the immediate future. Certain aspects of the situation which has arisen seem to call for watchfulness and, if necessary, quick action on the part of the Administration. I believe therefore that it is my duty to bring to your attention the following facts and recommendations:

In the course of the last four months it has been made probable - through the work of Joliot in France as well as Fermi and Szilard in America - that it may become possible to set up a nuclear chain reaction in a large mass of uranium, by which vast amounts of power and large quantities of new radium-like elements would be generated. Now it appears almost certain that this could be achieved in the immediate future.

This new phenomenon would also lead to the construction of bombs, and it is conceivable - though much less certain - that extremely powerful bombs of a new type may thus be constructed. A single bomb of this type, carried by boat and exploded in a port, might very well destroy the whole port together with some of the surrounding territory. However, such bombs might very well prove to be too heavy for transportation by air.

-2-

The United States has only very poor ores of uranium in moderate quantities. There is some good ore in Canada and the former Czechoslovakia, while the most important source of uranium is Belgian Congo.

In view of this situation you may think it desirable to have some permanent contact maintained between the Administration and the group of physicists working on chain reactions in America. One possible way of achieving this might be for you to entrust with this task a person who has your confidence and who could perhaps serve in an inofficial capacity. His task might comprise the following:

a) to approach Government Departments, keep them informed of the further development, and put forward recommendations for Government action, giving particular attention to the problem of securing a supply of uranium ore for the United States;

b) to speed up the experimental work, which is at present being carried on within the limits of the budgets of University laboratories, by providing funds, if such funds be required, through his contacts with private persons who are willing to make contributions for this cause, and perhaps also by obtaining the co-operation of industrial laboratories which have the necessary equipment.

I understand that Germany has actually stopped the sale of uranium from the Czechoslovakian mines which she has taken over. That she should have taken such early action might perhaps be understood on the ground that the son of the German Under-Secretary of State, von Weizsäcker, is attached to the Kaiser-Wilhelm-Institut in Berlin where some of the American work on uranium is now being repeated.

Yours very truly,

(Albert Einstein)

“A single bomb of this type, carried by boat and exploded in a port, might very well destroy the whole port together with some of the surrounding territory.”

In the massive explosion at Port Chicago, The SS *E.A. Bryan* "exploded as a single bomb".

As noted earlier, John von Neumann, the esteemed mathematician, was Einstein's colleague at the Institute for Advanced Studies; he was also a consultant to Los Alamos.

When von Neumann visited the Lab in September 1943, he and Parsons spent two days working behind closed doors. It appears that they were working on implosion, although implosion was not the priority at the time and did not take priority until July 1944.

In the fall of 1943, Parsons was interested in underwater delivery. It is reasonable to speculate that at least part of their conversation had to do with the effects of underwater delivery. Parsons consulted with von Neumann on that topic two years later, in preparation for *Operation Crossroads*. von Neumann calculated the effect of the shock wave and predicted that an underwater explosion would have significant force. As noted earlier, the original idea for an atomic bomb, as described in the so-called Einstein letter, was that

> "A single bomb of this type, carried by boat and exploded in a port, might very well destroy the whole port together with some of the surrounding territory."[150]

This idea corresponds to the conclusion by Rudolf Peierls and Otto Frisch, the two British physicists whose 1940 report spurred the American bomb project. Peierls and Frisch thought it would be an inappropriate weapon for use by Great Britain because of the unavoidable harm to civilians from radioactivity. Instead, they suggested that the bomb might be used as 'a depth charge near a naval base."[151]

The atomic bomb, then, was originally conceived of as a Naval weapon. That was the premise under which Parsons and the scientists and engineers working for the Manhattan Project ordnance division began their work on the *Thin Man* plutonium-gun model that was "shelved" on July 17th, 1944. During the course of research and development, they discovered that a smaller bomb could be created and that it could fit into a modified version of the new B-29 bombers. Meanwhile, Parsons' division studied the possibility of underwater delivery—perhaps via a nuclear torpedo.

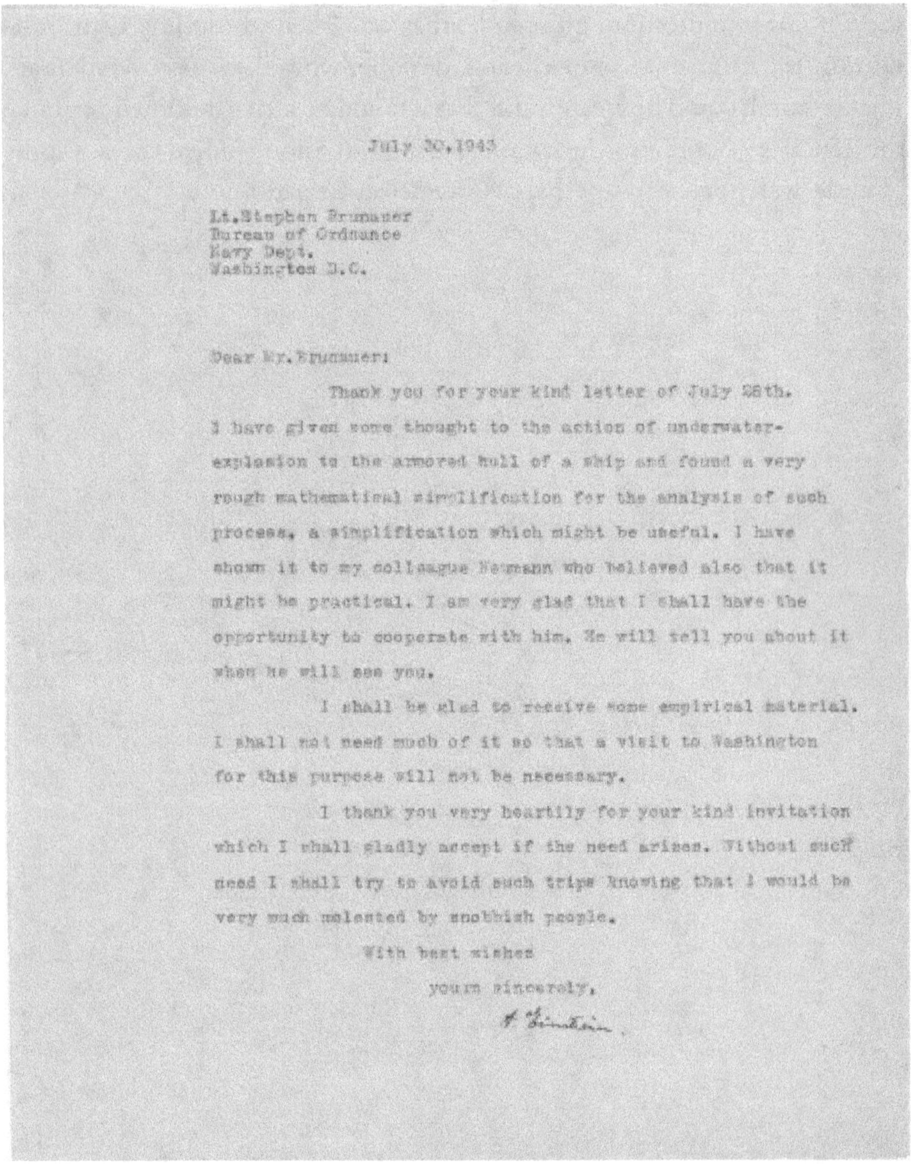

July 30, 1943

Lt. Stephen Brunauer
Bureau of Ordnance
Navy Dept.
Washington D.C.

Dear Mr. Brunauer:

Thank you for your kind letter of July 28th. I have given some thought to the action of underwater-explosion to the armored hull of a ship and found a very rough mathematical simplification for the analysis of such process, a simplification which might be useful. I have shown it to my colleague Neumann who believed also that it might be practical. I am very glad that I shall have the opportunity to cooperate with him. He will tell you about it when he will see you.

I shall be glad to receive some empirical material. I shall not need much of it so that a visit to Washington for this purpose will not be necessary.

I thank you very heartily for your kind invitation which I shall gladly accept if the need arises. Without such need I shall try to avoid such trips knowing that I would be very much molested by snobbish people.

With best wishes
yours sincerely,

"... I have given some thought to the action of underwater-explosion to the armored hull of a ship and found a very rough mathematical simplification for the analysis of such process ... I have shown it to my colleague, Neumann ... I am glad that I shall have the opportunity to cooperate with him...." - Einstein to Brunauer, July 20 1943

The Atomic Heritage organization makes the connection in an essay on the Navy's role in the post-war atomic program:

> Deak Parsons had *done research during the war on theoretical nuclear attacks on ships, including investigation into the 1944 Port Chicago disaster* in which 320 sailors were killed after munitions aboard a cargo ship detonated. In 1945, Parsons formed the Navy Atomic Bomb Group, and began to push for tests of bombs above and below ships. [152] [emphasis added]

It is not clear whether the author of this statement had some undisclosed insight into the Port Chicago explosion or simply drew the logical conclusion that since the head of ordnance at Los Alamos took time to investigate the scene, the Port Chicago explosion must have been of particular importance to the Project. There were explosions at other Naval bases, navy yards and ammunition depots between 1943 and 1945, but there is no indication that any of them held interest for Parsons or the Project.

One outstanding question must be answered in assessing the PCnet: even if the Project had produced a workable weapon by the summer of 1944, would it have been possible (or desirable) to test it at Port Chicago?

Parsons' power

In November 1943, while Parsons was still pursuing underwater delivery, Oppenheimer decided that method did not appear to be promising. He directed Parsons to focus on other methods, but it is not clear that Parsons followed his instructions. Christman points out, Parsons' connections limited the extent to which Oppenheimer and Groves exercised power over his actions.

In May 1943, after meeting with Groves to talk about the new job, Parsons visited Bush and Rear Admiral H. P." Spike", the two men Christman identifies as "the source of his power in the fuze program." (After the war, Blandy headed *Operation Crossroads*, the tests of atomic bombs against Naval ships for which Parsons was Technical Director.)

Meanwhile, Admiral H.R. Purnell, the Navy representative on the Military Policy Committee that advised Bush on the Project, decided to keep Parsons on the official staff of Admiral Ernest King.

> . . . Purnell and Parsons quickly recognized that there could be long-range advantages in keeping Parsons' name on the rolls of Admiral King's staff. For one thing, it provided security cover for Parsons, a plausible explanation of his whereabouts. When well-meaning friends saw him in Washington and wanted to know what he was doing, the answer was simple: "I'm on Admiral King's staff."[153]

Parsons traveled to Washington at least once a month, to meet with Groves, if for no other reason. But he no doubt accomplished a lot of other things while he was there. And he may have played a few hands of tennis with his friend, James Forrestal, Under Secretary of the Navy.

Parsons was connected to the top of both the Navy and the Manhattan Project. They trusted him to represent their interests. The scheme would be ideal, too, for arranging an experiment at Port Chicago. It is reasonable to *speculate* that the cover that allowed him to come and go as he pleased in Washington would apply as well to other places, including the ammunition dump (where he had stored material during the tests of the proximity fuze).

Parsons had access to whatever resources he needed as well, whether money, men or material. While working on the proximity fuze, he had commandeered a cruiser, the *USS Cleveland*, and requisitioned six radio-controlled aircraft to serve as pilotless drone targets. (He only got four.)

> More important, the phantom position gave him the authority to originate high-priority requests and purchases out of King's office without disclosing the Los Alamos connection. The ComlnCh position and the backing of Admiral Purnell added to Parsons' unique powers within Project Y. In short, Parsons went into the bomb project with a direct line to the summit of the U.S. Navy. He wore this power, as well as his influence with Vannevar Bush of the scientific community and Spike Blandy of BuOrd much like a concealed weapon. He knew it was there. Groves and Oppenheimer knew it was there. Project leaders at Los Alamos sensed its presence. Yet possessing the weapon, Parsons brandished or used it only when essential.[154]

This is not the portrait of a man whose claim to fame is the one act—however glamorous, heroic and unique—of arming the *Little Boy* bomb in midair. With that minimization of his role, which contributes to his obscurity, Parsons' cover extended across time as well. During the war emergency, he was able to do things that could not be done in peacetime.

FDR's big atomic nightstick

"What if the atomic bomb, which FDR knew by late 1944 was likely to be tested in the near future, would so intimidate Stalin with the image of overwhelming American power that he would relent on the core issues?

Campbell Craig,

The Atom Bomb as Policy Maker: FDR and the Road Not Taken

The Age of Hiroshima

Never in the history of humanity had there been a greater challenge, or a greater—and riskier—opportunity than the one facing Franklin D. Roosevelt at the onset of *World* War II.

It was a time of global upheaval. Science and technology were changing the world. Nations jockeyed for position as the promise of industrialization changed their prospects for status and power. With steady progress of military science and technology, Americans could no longer sit safely "across the pond", shaking their heads at the "phony war."

FDR's grand design envisioned the "civilized" world being governed by the "four policemen—the U.S., Great Britain, Russia and (to balance Russian influence) China. FDR went overboard to play nice with Joseph Stalin, whose cooperation was essential to his vision for the new world order that would be his legacy.[155]

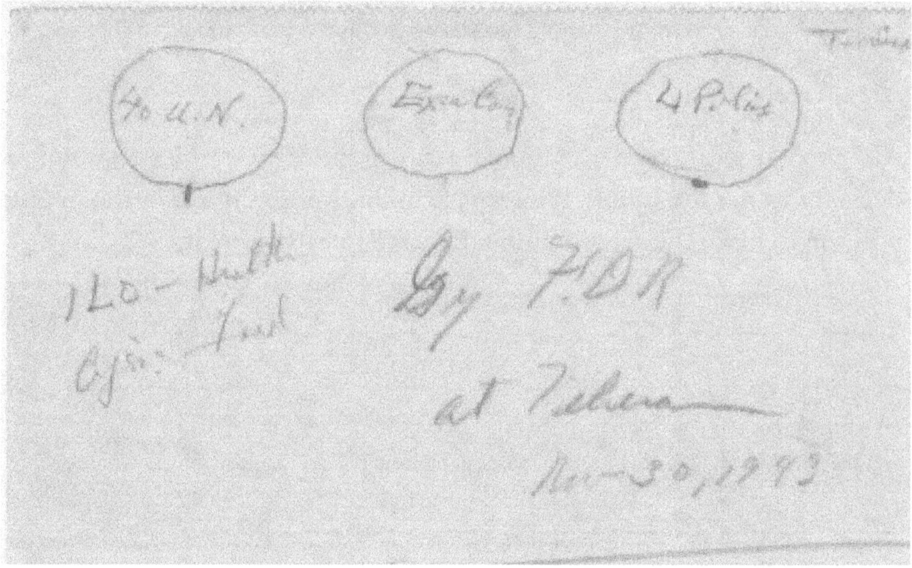

At the Tehran conference in November 1943, FDR sketched and signed and dated his idea for the "Four Policemen" to enforce the international laws of the United Nations.

FDR was a visionary internationalist whose great ambition was to establish a new world order through the creation of the United Nations. He considered stepping down from the Presidency to head the world-governing organization.

Although it was President Truman who presided over **the pre-weighted decision to *use* the bomb**, President Franklin Roosevelt was in the singular position to make **the secret, unilateral decision to *create* it.**

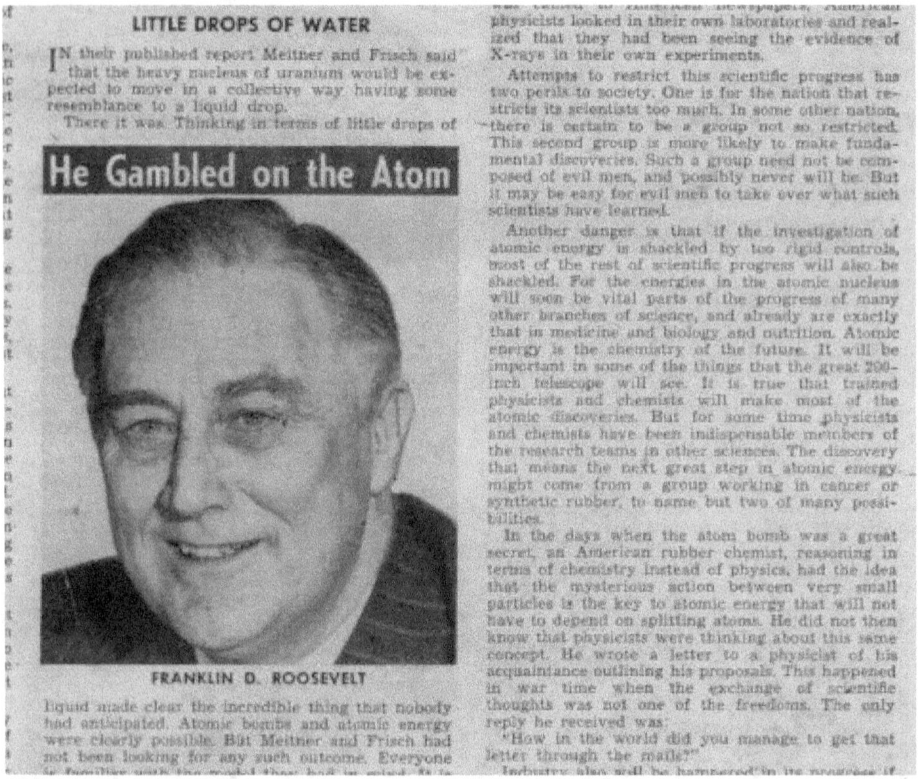

LITTLE DROPS OF WATER

IN their published report Meitner and Frisch said that the heavy nucleus of uranium would be expected to move in a collective way having some resemblance to a liquid drop.

There it was. Thinking in terms of little drops of

He Gambled on the Atom

FRANKLIN D. ROOSEVELT

liquid made clear the incredible thing that nobody had anticipated. Atomic bombs and atomic energy were clearly possible. But Meitner and Frisch had not been looking for any such outcome. Everyone

physicists looked in their own laboratories and realized that they had been seeing the evidence of X-rays in their own experiments.

Attempts to restrict this scientific progress has two perils to society. One is for the nation that restricts its scientists too much. In some other nation, there is certain to be a group not so restricted. This second group is more likely to make fundamental discoveries. Such a group need not be composed of evil men, and possibly never will be. But it may be easy for evil men to take over what such scientists have learned.

Another danger is that if the investigation of atomic energy is shackled by too rigid controls, most of the rest of scientific progress will also be shackled. For the energies in the atomic nucleus will soon be vital parts of the progress of many other branches of science, and already are exactly that in medicine and biology and nutrition. Atomic energy is the chemistry of the future. It will be important in some of the things that the great 200-inch telescope will see. It is true that trained physicists and chemists will make most of the atomic discoveries. But for some time physicists and chemists have been indispensable members of the research teams in other sciences. The discovery that means the next great step in atomic energy might come from a group working in cancer or synthetic rubber, to name but two of many possibilities.

In the days when the atom bomb was a great secret, an American rubber chemist, reasoning in terms of chemistry instead of physics, had the idea that the mysterious action between very small particles is the key to atomic energy that will not have to depend on splitting atoms. He did not then know that physicists were thinking about this same concept. He wrote a letter to a physicist of his acquaintance outlining his proposals. This happened in war time when the exchange of scientific thoughts was not one of the freedoms. The only reply he received was:

"How in the world did you manage to get that letter through the mails?"

"He gambled on the atom" - Howard Blakeslee, "The Atomic Future," **Associated Press** supplement, May 1946.[156]

The Truman administration, as it turned out, found it hard to share the new weapon with other nations. In 1945, as the United Nations prepared to address the delicate question of international control, the U.S. prepared its own plans for *Operation Crossroads*.

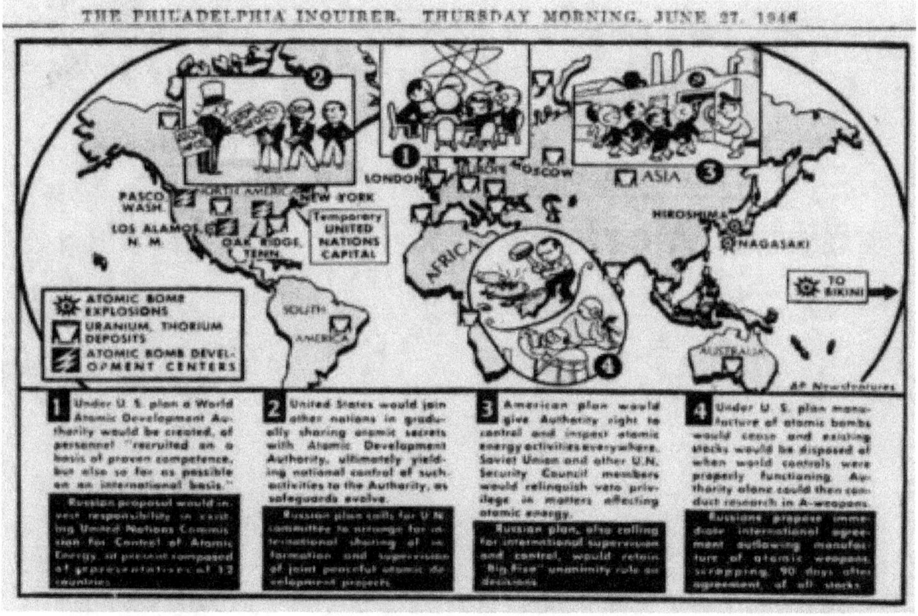

The Philadelphia Inquirer, June 27, 1946. - The U.S. simultaneously prepared for talks at the United Nations on international cooperation and the unilateral tests of atomic bombs against ships at *Operation Crossroads*.

As Admiral William H. P. Blandy put it, the world was at a crossroads. The American Century was well underway.

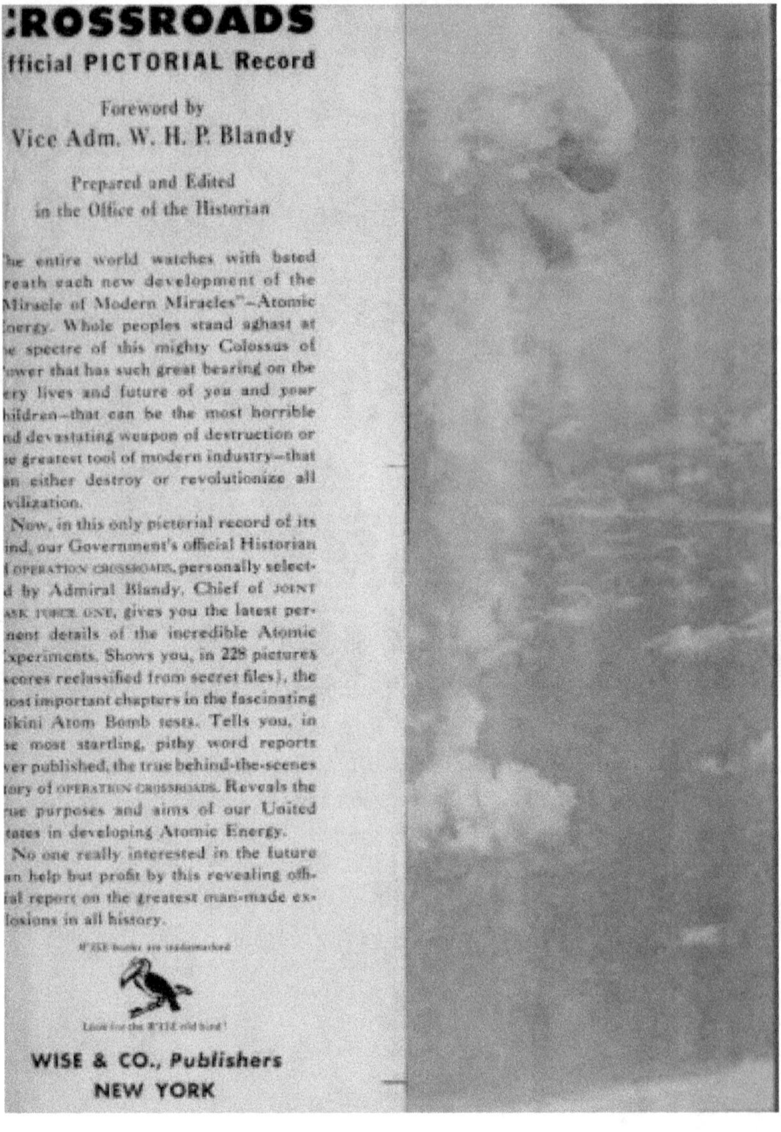

:ROSSROADS
fficial PICTORIAL Record

Foreword by
Vice Adm. W. H. P. Blandy

Prepared and Edited
in the Office of the Historian

he entire world watches with bated
reath each new development of the
Miracle of Modern Miracles"—Atomic
nergy. Whole peoples stand aghast at
e spectre of this mighty Colossus of
ower that has such great bearing on the
ery lives and future of you and your
hildren—that can be the most horrible
nd devastating weapon of destruction or
e greatest tool of modern industry—that
an either destroy or revolutionize all
ivilization.
Now, in this only pictorial record of its
ind, our Government's official Historian
OPERATION CROSSROADS, personally select-
d by Admiral Blandy, Chief of JOINT
ASK FORCE ONE, gives you the latest per-
nent details of the incredible Atomic
xperiments. Shows you, in 228 pictures
scores reclassified from secret files), the
ost important chapters in the fascinating
ikini Atom Bomb tests. Tells you, in
e most startling, pithy word reports
ver published, the true behind-the-scenes
tory of OPERATION CROSSROADS. Reveals the
ue purposes and aims of our United
tates in developing Atomic Energy.
No one really interested in the future
an help but profit by this revealing offi-
ial report on the greatest man-made ex-
losions in all history.

WISE & CO., Publishers
NEW YORK

"The entire world watches with bated breath each new development of the "Miracle of Modern Miracles"—Atomic Energy. Whole peoples stand aghast at the spectre of ***this mighty Colossus of power that has such great bearing on the very lives and future of you and your children****—that* ***can be the most horrible and devastating weapon of destruction*** *or the greatest tool of modern industry – that can either* ***destroy or revolutionize all civilization****."*

[emphasis added] - Vice Adm. W. H. P. Blandy, foreword, ***Operation Crossroads*: The Official Pictorial Record**, The Office of the Historian of U.S. Joint Task Force 1946

Race and the "race for the bomb"

Historians often shy away from it, for one reason and another, but racism was one form of supremacy that seasoned the values of world leaders in the war years. Not just in the U.S. but throughout the world, divisive policies like segregation, discrimination and colonialism compromised the welfare of peoples of color.

On the one hand, there was the ageless idea that some people were superior to others. Eugenics, the creation of a better human race through selective breeding, was popular, acceptable among diverse groups of Americans. The philosophy also influenced Hitler and his Aryanism. But other nations, too, categorized certain groups – including Jews, homosexuals, Roma, people with disfiguring disabilities and other vulnerable groups, including the dark-skinned races – as undesirable "others".

In **War Without Mercy,** John Dower makes it clear that this venomous "otherism" worked both ways, as exemplified in the beastly way Japan and America portrayed each other. Ignorance about "others" can breed stereotypes, which in turn breed fear, and fear – especially in times of crisis – breeds the hatred that accompanies intense nationalism.[157]

The distasteful, racist pictures in Dower's book, of course, are central to his message. But outright racial hatred is one way of expressing racism. Racial discrimination was also a pragmatic means to other ends. There was labor to be done, fields yet to be plowed and, with the war on, defense factories to be run.

Even 'inferior people' (like the "low quality personnel" who loaded ammunition at Port Chicago) could be useful in one way or another. As FDR pointed out in the summer of July 1942, shortly after approving the Manhattan Project, minorities were vital to victory.

And, of course, many people of "superior cast" believed they had the right and responsibility to manage the affairs of the world. If they benefited from their status and contributions, that too was in order.

The fictitious element "*vibranium*" in the marvelous world of *Wakanda*, the secret African country in the Black Panther movie, had its real counterpart in the great **uranium** deposits mined by Black African laborers in the Shinkolobwe mine in the Haut-**Katanga** Province of the Democratic Republic of the Congo. This is the uranium that made the American bomb – and to a great extent, the American Century – possible.

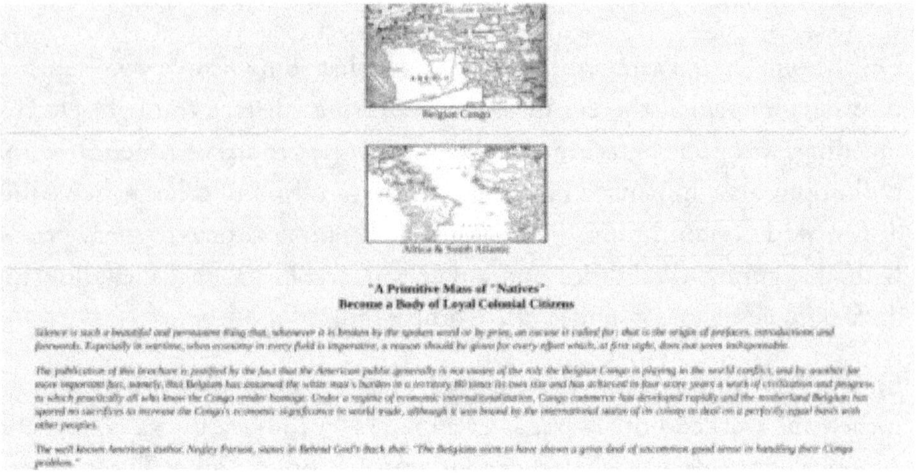

"*. . . The publication of this brochure is justified by the fact that the American public generally is not aware of the role the Belgian Congo is playing in the world conflict, and by another far more important fact, namely, that Belgium has assumed the white man's burden . . .* "[158]

Economic and social realities are mutually influential. Furthermore, individual attitudes are sometimes influenced, if not shaped, by national values, and individual actions by national actions. The reverse is also true. With regards to race, a social climate of racial hostility undergirds racist acts on the part of government leaders, in America and throughout the world.

It is never pleasant for anyone to think about racism from any perspective. That makes it all the more imperative to understand if, when and how racism was a factor in world-shaping events like the creation and use of the atomic bomb.

Racism was beyond doubt one aspect of the world's wars. To overlook or downplay this integral part of history is to miss a crucial element that helps explain the past so we can live better in the future.

The Port Chicago nuclear explosion theory must be seen in this large, world-war size context; for this is the context in which FDR presided over America's rise to superpower status.

What if...

The hidden history of the bomb contains clues to several puzzling issues that confound the contributors to **The Age of Hiroshima**, including questions about nuclear proliferation, nuclear deterrence, and the puzzling issue of why there has been no further use of the bomb *in combat* since 1945.

In **The Age of Hiroshima**, Campbell Craig explores questions about how the bomb may have impacted FDR's plans for the post-world order, and for his legacy ambition, the United Nations. Campbell's essay, "*The Atom Bomb as Policy Maker: FDR and the Road Not Taken*"[159], poses an unusual "what if" question:

> "What if the atomic bomb, which FDR knew by late 1944 was likely to be tested in the near future, would so intimidate Stalin with the image of overwhelming American power that he would relent on the core issues? Could the simple prospect of overarching American economic might, now combined with a revolutionary weapon, persuade Stalin to accede to Roosevelt's security order?"[160]

Otherwise stated, atomic diplomacy may have been FDR's idea, one that Truman inherited with the bomb. FDR may have adhered to the philosophy of his uncle, mentor and predecessor, Theodore Roosevelt, who expressed his foreign policy in the words of an old African proverb: "Speak softly and carry a big stick. You will go far in the world."

"Our bomb" was in the hands of the good guys, those who would only use it for good – according to those who used it.

Ever since the bombing of Hiroshima, when the public first became aware of the bomb, some people have seen it as a gift from God. Truman said as much: thank God we got it first.

This widespread and *persistent* sentiment reflects FDR's conclusion, as recorded by Sachs, that "... pursuit of a nuclear bomb might be a providential challenge and opportunity for Americans and democracy."

Though the fact is seldom recognized, the top-secret opportunity for America was indeed a challenge to American democracy. In a post-war speech before the Naval War College, Parsons made the admittedly shocking declaration that the U.S. had been under a dictatorship during the war.[161]

As Michael Uhl and Tod Ensign put it in **GI Guinea Pigs: How the Pentagon Exposed Our Troops to Dangers More Deadly Than War**

> The world's first fission bomb was built in absolute secrecy by the Manhattan Project in the midst of World War II, and strict secrecy continued to be the policy followed by post-war nuclear test program until well into the 1950s. **The fact that the basic policy decisions about the development and use of nuclear power were made by a closely knit circle of scientists and public officials, insulated from public pressure and opinion, was to have enormous consequences."** [emphasis added]

The head of ordnance for the Manhattan Project was a former aide to Bush, who had the President's ear. Each of these three powerful, secretive and visionary men understood the occasional need to be ruthless.

Whether or not we agree, these men spoke for all Americans. Whether or not we approve of what they said and did, they acted on our behalf.

Now we know.

Evidence pertaining to the PCnet, a review

An investigation of the Port Chicago nuclear theory requires research into a variety of topics. The research unearths a wide-ranging body of previously unexplored evidence.

This information updates early nuclear history. It provides perspective for the current status of nuclear policy. And, taken together, this little-known evidence supports the hypothesis that Captain William S. Parsons, presumably acting under the authority of Vannevar Bush and on behalf of President Franklin D. Roosevelt, may have staged a secret nuclear detonation resulting in the Port Chicago explosion of July 17, 1944.

To recap:

The Port Chicago Naval Ammunition Depot provided a good camouflage as well as a convenient site for a secret test.

> • Although the PCNAD was supposedly an important facility for shipment of ammunition to the Pacific, the hazardous conditions there were well-known and avoidable, yet went unaddressed. The site was considered "an accident waiting to happen." Through the lens of the PCnet, it may have been an accident *designed to appear* to have happened.
>
> • The timeline of the PCNAD corresponds with the development of the Manhattan Project. For example:

- Ground was broken on the PCNAD around the same time the Navy began admitting Negroes to all ratings;
- The Navy began admitting Negroes in April 1942, at the same time Bush was developing plans for the Project.
- The first Negro sailors were available to load ammunition at the base when it opened in December 1942.
- The criteria that made the base a good location for a transshipment facility—specifically, a remote port, under military control—were

also the criteria for a nuclear proving ground

The Port Chicago Disaster made a good cover for a nuclear explosion.

- The story of one of the greatest explosions in history was soon overshadowed by the jerry-rigged Port Chicago court martial, which was:

- a mass court martial, the largest in Naval history
- publicized by the Navy and open to the participation of the NAACP;
- based on a charge of conspiracy to commit mutiny, which was questionable even at the time, especially since

▪ there was no attempt to usurp authority and no violence;

▪ according to Naval regulations (in the **Bluejacket Manual**, issued to every new enlistee) the sailors could have been disciplined individually, without need of any tribunal, for refusing to obey an order;

▪ the "work stoppage" at Port Chicago was a relatively minor event, especially in comparison to riots and disturbances at other bases, including nearby Vallejo, during the war;[162]

▪ the charges were immediately reduced after the war and most of the sailors were restored to service, and ultimately discharged with honor from the Navy. Yet the charge stayed on their records.

▪ the mutiny charge is still being contested eighty years later.

- The story of the Port Chicago *explosion* is hardly known, although

- it was "the greatest home-front disaster of World War II";
- it is still listed as one of the largest man-made explosions in history;

- it had the same magnitude of the atomic bomb that was dropped on Hiroshima;
- was studied as part of the research and development of the atomic bomb, providing essential data that was otherwise unavailable.

• The Court of Inquiry that investigated the explosion concluded that the cause could never be known; yet

- most accounts emphasize the dramatic scenario that perpetuates the *inaccurate assumption* that the explosion was caused by mishandling of explosive ammunition by untrained Black sailors working under hazardous conditions and supervised by inexperienced and racist White officers;
- one of the men on the three-man court, Captain James Crenshaw, was the brother-in-law of the man who led the Ordnance Division of the Manhattan Project, Captain William S. Parsons, the man most apt to have conducted a nuclear test at Port Chicago. Having a trusted friend on the Board would have been convenient for a cover up.

<u>The History of the Bomb</u>

• The history of the bomb is full of gaps and unanswered questions; many of the answers pertain to the PCnet in one way or another.

• One of the reasons for the faulty history is that it built upon the deliberately deceptive postwar story promulgated by Manhattan Project leaders and government officials who wanted to sell the bomb to the public and avoid Congressional hearings.

• The Einstein letter story that introduces almost every history of the bomb was a deliberate fabrication; it was designed to sell the bomb by associating it with the beloved genius scientist.

• Fear of a German bomb motivated many of the *emigrant* scientists who were involved with the bomb project, but they had little influence on its creation and less influence on its use. American scientists had other motivations, including patriotism, of course; the opportunity to use their skills to make a contribution to the war effort (without military service); the singular opportunity to work on the "sweet" cutting edge science, with brilliant colleagues from around the world, and with unlimited funding.

• The motivating factors most discussed among key people involved *in the decision to create the bomb* was not fear of a German bomb, nor even the need to win the war and save lives; but rather the military advantage of being the first nation to get the unprecedented weapon.

• The Manhattan Project became the very icon of a successful government undertaking despite the fact that

- it was kept secret (*primarily*) from the American public;
- it resulted from a unilateral decision made single-handedly by the President and was administered by a new agency set up under FDR's Executive branch without Congressional oversight
- it operated on a black budget
- it had all the other earmarks of a conspiracy;
- it resulted in the creation of a weapon of mass destruction that still plagues the world today.

<u>President Franklin D. Roosevelt</u>

• FDR was the president who unilaterally authorized the American bomb project; yet his name is barely associated with it.

• His motives for creating the bomb provide a key rationale and are essential to an accurate history; yet historians have only recently begun to explore his role in the decision to create it and to use it.

- FDR was a visionary internationalist whose push for the United Nations would not only create a new world order, but would establish his legacy and possibly even allow him to become the first head of a one-world government.
- Even before its conception, the greatest (and most obvious but least explored) use for the "exponential" weapon was its built-in ability to influence international affairs.
- Evidence of FDR's thoughts and plans for the weapon are detailed in the papers of his advisor, Alexander Sachs; but Sachs has been relegated to a minor role and his testimony has been deemed unreliable.

● Journalists and historians have been unable to establish a single compelling motive for FDR's sudden secret trip to the West Coast in July 1944.

● FDR was on the West Coast at the time of the Port Chicago explosion.

● FDR's flagship for the Honolulu Conference, the USS *Baltimore* (CA-68), was anchored at the nearby Mare Island Navy Shipyard, along with five other ships in the fleet, from July 6th to July 17th, the day of the explosion. No journalists or historians have taken note of this fact; and the COI ruled out the likelihood of sabotage at Port Chicago.

● Instead of a ship that was already stateside, the *Baltimore* had been called from the South Pacific for the special assignment; while hastening to the U.S., the unescorted ship made unexplained stops in two locations. (Whether or not they picked up passengers would be a subject for further research; one place to begin would be to identify the whereabouts of General Douglass

MacArthur and Admiral Chester Nimitz during the month prior to the Honolulu Conference.)

• The President's ship was allegedly sent to Mare Island for remodeling to accommodate his trip; yet he embarked for Hawaii at San Diego, where the Navy Shipyard was a designated ship repair facility.

• The Democratic National Convention was taking place in Chicago at the same time. Nothing about the Honolulu Conference explains why the President missed the historic opportunity to accept his unprecedented fourth nomination in person.

• There is no specific account of the whereabouts of FDR's train, or, more importantly, of his own whereabouts, between 5:00 p.m. on July 17th and the wee hours of July 19th, when his ship appeared at San Diego, without the usual fanfare.

• FDR was prone to intrigue; he had been known to appear in one place when he was alleged to be in another. He also liked to see things for himself, and had gone out of his way to witness other military operations.

• On the morning of July 20th, hours before he gave his acceptance speech, FDR had a sudden attack of pain, possibly a heart attack. (In the spring, his doctor had been concerned that he not be told about the state of his heart because the bad news might trigger an adverse reaction.) His health deteriorated rapidly from that point on.

• Several people, including some of the generals and one serviceman at Honolulu, took note of FDR's ghastly appearance (which was comparable to the way Mrs. James Conant described her husband's appearance and demeanor after witnessing the

Trinity test. The generals found the President's appearance all the more shocking because he had just taken a cross-country train trip followed by the leisurely cruise to Hawaii. His first speech upon returning to the U.S. was noted as his worst; he rambled and stumbled.

President Harry Truman

● Truman is the president most associated with the bomb, though he had little to do with the decision to use it, especially since

- decisions leading up to its use had already made; including potential targets and delivery methods;
- the extreme investment in the secret bomb project would have made it subject to unrelenting Congressional scrutiny if the bomb had not been used;
- preparations for its use began long before Truman became President;
- his advisors were men from FDR's administration, who were already committed to using the bomb.

● Although the "accidental president" wound up presiding over one of the major decisions of the twentieth century, the upheaval at the Democratic National Convention when Truman was nominated has come under little scrutiny.

● Historians insist that Truman knew nothing about the bomb until after his inauguration, despite clear evidence that he did know and very little to substantiate the claim that he did not.

Alexander Sachs

● By all accounts, Sachs, FDR's economic adviser, was the first person to introduce him to the bomb; yet this significant historical fact is barely known.

• In several instances, Sachs was the only man in the room, and his papers are the only source of information about conversations he had with FDR. Yet these unique primary documents have been used selectively and otherwise largely ignored by historians.

• Sachs made several other significant contributions to the bomb project—he participated on the Uranium Committee, for example, and facilitated the import of uranium—yet he is best and almost exclusively known simply as FDR's friend / advisor who delivered the "Einstein" letter.

• Sachs made several claims that counter the flawed Einstein letter story, yet historians have overlooked his claims without comment.

<u>The Navy, Parsons, and Albert Einstein</u>

• The Navy was allegedly left out of the bomb project, although

- There are conflicting explanations of who made the decision and why.
- Each of the explanations is dubious.

• The Navy made several contributions to the project, including the participation of key Naval personnel and transportation of the bomb to the Pacific

• Omission of the Navy's role in the Project helps to conceal several clues to the PCnet, including

• the role of Captain William S. Parsons, who would have been the man responsible for a nuclear test at Port Chicago;

• the fact that Parsons pursued underwater delivery of the bomb—i.e., a nuclear torpedo—for the first half of the Project;

• the fact that during that same period, Einstein (who was also an inventor) worked on torpedo design as a consultant for the Navy;

• the fact that after the war, one of Einstein's Navy liaisons, Lt. Stephen Brunauer, was Officer-in-Charge at *Operation Crossroads*, the 1946 tests of the atomic bomb against Naval ships, serving under Parsons, who was Technical Director.

Einstein

• Einstein immigrated to the U.S. in 1933; the Institute for Advanced Studies at Princeton may have been set up largely to accommodate him.

• In January 1934, the Einstein's were overnight visitors at the White House; yet few people know about the extended visit between Einstein and FDR, two of the most important figures of the twentieth century, who each had a little-known role in the creation of the atomic bomb.

• As indicated in the so-called Einstein letter, the first idea for a bomb was that it would be transported by ship, and that was the premise when work on the weapon began;

• The original plan calling for a naval weapon explains in part why the head of Ordnance for the Project was a naval officer;

• Reference in the Einstein letter to "a single bomb exploding in a port" matches the description of what happened to one of the ships at Port Chicago. The SS *E.A. Bryan* exploded like a single bomb.

• At one point in his work on torpedo design, Einstein indicated that an experiment was necessary; theoretical work could not resolve the issues under study.

● Einstein recommended two explosions; there were two explosions at Port Chicago

Parsons, Bush, and the Manhattan Project

● Parsons is best known for a relatively minor role in comparison to his major contributions to the war and to history. He remains an obscure historical figure although he

- was the man responsible for the creation and delivery of the bomb;
- was well-connected and had unprecedented autonomy and power even before joining the Project;
- was instrumental in the development of the two other key military technologies of World War II;
- became an Associate Director of the Los Alamos Laboratory;
- achieved the rank of Rear Admiral—the first "atomic admiral"—without commanding a ship;
- was instrumental in the development of the first continental nuclear proving grounds (the Nevada Proving Grounds);
- helped establish permanent U.S. nuclear policy;
- hired several key personnel who worked for the Project;
- had personal and professional ties to many top scientists and military leaders;
- died suddenly of a heart attack at the age of 53, allegedly triggered by hearing the bad news that Oppenheimer's security status was *going to be* suspended. (The PCnet speculates that if there was a link between Parsons' death and Oppenheimer's status, it may have been connected to the fact that Oppenheimer insisted on a hearing. In that case, it may shed light on what the Lab director meant when he told President Truman, "I'm afraid we have blood on our hands.")

● Before joining the Project, Parsons was Special Assistant to Bush.

Vannevar Bush

- As the head of the Office of Scientific Research and Development (OSRD), Bush reported directly to FDR; it was Bush who coordinated the research that led to the Manhattan Project and recommended it to FDR.

- Bush was involved with the earliest decisions about the bomb; Parsons, a leading expert on explosives and military experiments, was his assistant during that time. In his memoir, *Pieces of the Action*, Bush offers no names but states that in advising FDR on matters beyond his expertise, he often consulted with those who were experts in their fields.

- Bush recommended Parsons to head the Ordnance Division of the Manhattan Project.

- Bush administered both the Manhattan Project and the Committee on Medical Research (CMR), which was the agency that sponsored the Tuskegee syphilis study.

- From the beginning, Bush was aware of the eventual need for a remote location to test the weapon. (It is reasonable to believe he would have consulted with Parsons, the ordnance expert.)

<u>Parsons, the Manhattan Project and the Port Chicago Explosion</u>

- On the very day of the explosion, leaders of the Manhattan Project met in Port Chicago to decide the fate of the *Thin Man* plutonium gun-assembly bomb. The *Thin Man* had been Parsons' priority for over a year, yet he was absent from the meeting, and there was apparently no explanation given.

- Parsons had sent a memo to Groves indicating that he expected to be on the West Coast beginning June 12, 1944. In a September 1944 memo, he used the phrase, "From my observation of the Port Chicago explosion . . ." (He expressed disappointment in the effects of the detonation.)

• Parsons suggested that Groves should invite the head of the Fourth Army Air Force to observe one of the experiments on the West Coast as a means of courting his cooperation and some "rapid action" that was to be needed.

• The possibility that plutonium would be ineffective in a gun-assembly bomb was discussed as early as April 1943, but the *Thin Man* was the first priority project – the prototype atomic bomb – because it would be proof of the uranium gun. As such, it did not need to be a sturdy weapon; all that was needed was one demonstration that it would work.

• Parsons visited the site of the explosion in person, to study "the effects of the detonation," within two days of the explosion (and possibly sooner.)

• In his first memo after visiting the site, Parsons said ". . . Port Chicago was designed for large explosions." (Perhaps if he could have been more open – about a top-secret, coded, clandestine operation, largely carried out by verbal "mission orders" – Parsons might have substituted the word *atomic* for *large*. A literal interpretation of his statement could mean that the PCNAD was designed so that scientists and engineers developing the atomic bomb could ascertain just what type and level of damages were to be expected from an unprecedented nuclear detonation.)

• Parsons, in a postwar speech to the Naval War College, said the data from Port Chicago provided the first realistic expectation of the effects of a nuclear explosion.[163]

There was more to gain than to lose in a proof-test of the *Thin Man* at Port Chicago. The casualties and damage to property were part of the costs of the test no less than combat casualties were the cost of war. If Port Chicago was literally "designed for large explosions," as Parsons stated—if the depot was designed to be a proving ground for the atomic bomb—then failure to test the bomb there would have been the same as not using the atomic bomb in the war. In that case, the huge waste of money and other resources would have been the subject of never-ending Congressional hearings.

Ideally, the experiment would disprove the scientists' theoretical predictions that plutonium would predetonate in a gun-assembly weapon. In any case, the test would yield information that would be useful in designing the uranium gun.

Parsons was determined to let nothing prevent the perfect delivery of the perfect weapon. He, too, got what he wanted. "Once in a century," he said, "you can't shake off the Midas touch. That's what happened to us." His touch, though, was a different form of gold.

The end of the American century

The nuclear age is a by-product of the American Century. President Kennedy referred to the threat of nuclear war as the "Sword of Damocles", a dangerous weapon suspended above the world. Historians considering the "second nuclear age" may be chasing a ghost with a butterfly net. Why, they ponder, has there been no use of nuclear weapons since 1945?

Ellsberg dispels that cliché.

> "U.S. Presidents *have used* our nuclear weapons dozens of times in "crises", mostly in secret from the American public (though not from its adversaries). They have used them in the precise way that a gun is used when it is pointed at someone in a confrontation, whether or not the trigger is pulled. To get one's way without pulling the trigger is a major purpose for owning the gun."[164] [*original* emphasis]

The gun points both ways, leaking economic and environmental and psychological fallout.

It is time for a new world order, based on the premise that now, to complement our vast scientific knowledge and technical prowess, we need a different, kinder touch. That is to say, **what the world needs now is love**. And the normalization of uncommon sense.

As the American Century closes, the 21st century must become the People's Century, just as Vice President Henry Wallace proposed at the beginning of World War II.

There is hope.

The past is a mirror for us to look into when we want to correct the flaws in our national character.

On June 17, 2021, The National Academy of Sciences (NAS)—the agency that recommended the production of the bomb in 1941 – released a report, *The Future of U.S. Nuclear Weapons Policy,* in which they recommend major revisions.

Even the most hawkish military officers give lip service to a nuclear-free world; but to eliminate all nuclear weapons would be more revolutionary than burning bras, or flags, or even crosses. And there are some risks involved. But the work that must be done can be done. As the NAS report says,

> "...there have been, and continue to be, profound changes in the structure of the international order that are acting to reduce the probability of major war independent of nuclear deterrence. Moreover, even if all nuclear weapons were eliminated, the inherent capacities to rebuild them could act as a deterrent to the outbreak of major wars."[165]

What if that idea were to 'trickle up'? That would be a start. And maybe we could use some of the money spent on nuclear weapons to relieve the ongoing suffering of the nuclear survivors and their families. And to promote world peace. And to relieve poverty... Oh, how idealistic it sounds, even nutty. But why is it more difficult to feed hungry people than it is to split the atom and harness the energy of the sun?

Meanwhile . . .

- The man who presided over the U.S. with his finger on the button at the dawn of the second nuclear age borrowed language from President Harry Truman, who threatened a rain of ruin if Japan did not accede to Allied demands.

- Scholars published books and articles on the new challenges of the second nuclear revolution, as the U.S. Air Force contracted with Northrop-Grumman to build a new intercontinental ballistic missile.

According to the *Bulletin of the Atomic Scientists* (founded after the war by Albert Einstein and other scientists, newly "woke" by the alarm of nuclear war), the new missile will be

". . . the length of a bowling lane. It will be able to travel some 6,000 miles, carrying a warhead more than 20 times more powerful than the atomic bomb dropped on Hiroshima. It will be able to kill hundreds of thousands of people in a single shot."[166]

• The Russian invasion of Ukraine brought concerns about nuclear instability back to the headlines. While China and the U.S. circle Taiwan. (These, of course, are understatements about affairs that are ongoing at the time this book undergoes its final edit.)

• the madness escalates. Nuclear weapons have come a long way over the past eight decades; nuclear policy has stood its ground on the need for mutually assured deterrence.

Unless something changes by that time, the new missile will defend an old policy. Defense Secretary Dr. Mark Esper explained that modernizing our nuclear weapons is "key to our nation's defense. It provides that strategic nuclear deterrent that we depend on day after day – that we've depended on decade after decade."

In June 1946, as the newly formed United Nations gathered to consider the Baruch plan for international control of nuclear weapons, and as our military forces set the stage for *Operation Crossroads*, the grand-scale tests of nuclear weapons against naval ships, Lewis Mumford, envisioning this very state of affairs, said (no doubt shaking his head), "We in America are living among madmen."[167]

Meanwhile, as of January 2022, the Doomsday clock still sits at 100 seconds to midnight.

It is certainly time for a change.

Change will come no easier in the twenty-first century than it did in the twentieth; but it will come.

Because the small group of powerful White men who ran the world in the last century had different values, a different vision and a different worldview, than the *critical masses* around the world today who want and are willing to work for peace, in ways both as large as a nuclear explosion and as small as the mighty atom.

If we believe the **Washington Post** slogan, "Democracy Dies in Darkness", then we must realize that our democracy took a big hit during the war, when, according to W. S. Parsons, Russia's post-war dictatorship was "more rigid and perhaps more dynamic than the one we had operated under during World War Two."[168]

In the waning "American century", during this time of global upheaval when science and technology continue to bring us the mixed blessings of Western prosperity, perhaps it is time to reconsider the proposition of FDR's eclectic Vice President, Henry Wallace, and declare this the Century of the People.

Verdict?

Did the FDR administration authorize leaders of the Manhattan Project to conduct a secret nuclear test at Port Chicago on July 17th, 1944? Maybe.

If the PCnet teaches us nothing else, it tells us that there are secrets yet to be discovered about the history of the bomb.

This emerging history underscores the basic democratic principle that the watchfulness of American citizens is necessary to prevent government leaders and other powerful people from carrying out and covering up unacceptable acts that may have negative long-term impact. (It should cause us to reconsider our values and decide to what extent and how we celebrate the success of the Manhattan Project.)

Above all, exploration of the Port Chicago nuclear explosion theory should remind us that the bell of injustice tolls for each and all of us. Whenever anyone's life – including the lives of dehumanized "others" – can be secretly trumped by the ambitions of a few powerful people, what it ultimately means is that no one's life matters.

We can no longer leave the critical decisions about how our world is run to the ambitions and inclinations of a small group with unchecked power. We need the involvement of the world's ordinary people. I believe that we, the 'wee' people of the entire world can and must now engage in a chain reaction of life-supporting behavior and policies. For the real good of all humanity.

But where, you might ask, do we start?

With the freedom-loving, freedom-serving truth.

Acknowledgements:

I acknowledge that I believe in God; and I acknowledge that without God's guidance and help, I would have been both unwilling and unable to conduct the research and finish this book, against many incredible and otherwise unbearable odds. To me, both my very life and the completion of this book bear testimony to God's grace, which is able to continually make VIPs of me and you.

(Of course, *my perception* of Divine Providence should be a separate matter from *your evaluation* of the evidence presented in this book. In both cases, though, *knowing* the truth is the key to freedom.)

Thanks: To Ev, for helping with the book cover, for listening and asking questions, and for the other ways you supported and encouraged me. To Lyn, for helping me with my files, listening, reading, and the other ways you supported and encouraged me. To Alverda; Shirley J; Aleta J; Mary Y; Bob M; Mary D; Dr. Cassie; Shirley L and Diva; JC; Davie; and **everyone else** who listened, challenged, believed (or provided their challenging doubt); and especially to **all those who prayed me through**.

To **Lee Basham**, whose support and guidance has been nothing short of empowering. I too am grateful to share this 'lonesome journey', and look forward to the widening of the road.

Above all, speaking as a citizen, on behalf of our nation, I offer my **special thanks to Peter Vogel**. People, we owe him big time.

Dedication

For my oldest granddaughter: MaKiya, may your stint in the Navy be a blessed and honorable experience, and may your country deserve your service.

This book is dedicated to each and all of my six grandchildren, Jon, MaKiya, Autumn, Melody, Mykinzie and Harmony; and all the members of their generation; and for generations to come.

For the families of the victims and survivors of the Port Chicago explosion, and all the victims and families in the U.S. and throughout the world who have suffered so severely from the various hazards of nuclear weapons;

and for peace-loving Victors-In-Progress everywhere.

#EachAndAll4EachAndAll

Selected sources:

Most of the sources I used to research the PCnet and related topics, including some of the books and nearly all other sources listed below, were accessed digitally.

Websites:

The following websites are among the many invaluable sources for researching the various subtopics related to the PCnet:

- **atomicarchive.com**

- **Department of Energy** (DOE) OpenNet documents (osti.gov)

- **FDR Presidential Library & Museum** / FDR Museum: The Atomic Bomb

- **Harry S. Truman Library** (trumanlibrary.gov)

- **Fold3** - *Historical military records* (includes *Port Chicago War Diary*, *Mare Island War Diary*, and other military records)

- **Library of Congress online catalog**

- **The Manhattan Project: An Interactive History** (osti.gov)

- **National Archives online catalog**

- **The Nuclear Vault** - National Security Archive, George Washington University

- **The Nuclear Weapon Archive** - A Guide to Nuclear Weapons

- **Restricted Data** (The Nuclear Secrecy Blog) (Alex Wellerstein)

- **HyperWar Project** - Patrick Clancey, December 26, 2001; Currently maintained by Otto Torriero; Updated: 27 January 2019

- **Wikipedia**: *Talk:Port Chicago disaster*/Archive 1

Primary sources: (loosely defined here as material produced before, during or immediately after the war)

A variety of primary sources were used in the research of the PCnet. *Many* are identified in the context of this book, as well as *in my other writing* on the subject. The following links access ***some*** of the *primary resources* that are most critical for understanding the PCnet.

- Alexander Sachs papers; FDR library

- *The Evaluation of the Atomic Bomb as a Military Weapon*, Joint Chiefs of Staff, Harry S. Truman Library, Folder: "The Evaluation of the Atomic Bomb as a Military Weapon" Collection: The Decision to Drop the Atomic Bomb Series: President's Secretary's File

- *The Atom Bomb as Policy Maker* by Bernard Brodie, Foreign Affairs, Vol. 27, No. 1 (Oct., 1948), pp. 17-33 (17 pages): Council on Foreign Relations

- **National Archives:** *letters received from Albert Einstein (*to Lt. Stephen Brunauer)

- *The Atomic Future* - Howard W. Blakeslee, **Associated Press** supplement, May 1946, from **Alexander Sachs papers**

- *A Story Too Good to Kill: The "Nuclear" Explosion in San Francisco Bay*, by Lawrence Badash and Richard G. Hewlett, First Published June 1, 1993 **Sage Knowledge**

- *Background and Early History, Atomic Bomb Project in Relation to President Roosevelt, Opening Testimony by Alexander Sachs*, **Senate Special Committee on Atomic Energy**, Nov. 27, 1945

- *Capabilities of the Atomic Bomb, Including Naval Thinking on its Employment* - W. S. Parsons, **Naval War College Review**, April 1950

- *FDR Meets with Black Leaders.* Side 1, 1637-1972. September 27, 1940. **Transcripts of White House Office Conversations**, 08/22/1940 - 10/10/1940

- *Findings of Facts, Opinion, and Recommendations,* **Court of Inquiry, appointed by the Commandant of the Twelfth Naval District** to Investigate the Facts Surrounding The Explosion of 17 July 1944

- *AN HISTORIC VOYAGE WITH PRESIDENT OF THE UNITED STATES, FRANKLIN DELANO ROOSEVELT.* By Douglas E. MacVane at **Forward with Roosevelt**, the blog of the Franklin D. Roosevelt Library and Museum

- *Memorandum on Port Chicago Disaster*, W.S. Parsons, atomicarchive.com

- *Trinity test press releases* (May 1945) (Courtesy of Alex Wellerstein, **Nuclear Secrecy blog**)

<u>Secondary sources</u>: (loosely defined here as material *not* produced before, during or immediately after the war)

Research of the PCnet involved review of a wide range of sources, covering a variety of subtopics. **Many** of the following secondary sources are cited in the context of this book. **Some** of those select titles are included in the following list, which reflects those most essential to understanding the origin and significance of the theory. This is by no means an exhaustive list. More comprehensive resources will be provided in future books, and many sources can be accessed through The Port Chicago Witness Facebook page, where I post related articles as I see them, generally without comment.

Books / book chapters

The Age of Hiroshima, Princeton University Press (Jan 14, 2020); edited by Michael D. Gordin and G. John Ikenberry; ***see especially*** essays by Campbell Craig; Alex Wellerstein; Sean L. Malloy, and Francis J. Gavin

- *The Atom Bomb as Policy Maker: FDR and the Road Not Taken*, by Campbell Craig, **The Age of Hiroshima**, p 34

- *The Kyoto Misconception: WHAT TRUMAN KNEW, AND DIDN'T KNOW, ABOUT HIROSHIMA*, by Alex Wellerstein, **The Age of Hiroshima** (pp. 34-55)

- *"When You Have to Deal with a Beast": RACE, IDEOLOGY, AND THE DECISION TO USE THE ATOMIC BOMB,* by Sean L. Malloy, **The Age of Hiroshima** (pp. 56-70)

- *History and the Unanswered Questions of the Nuclear Age: REFLECTIONS ON ASSUMPTIONS, UNCERTAINTY, AND METHOD IN NUCLEAR STUDIES*, by Francis J. Gavin, **The Age of Hiroshima,** (pp. 294-312)

Albert Einstein: Creator and Rebel, by Banesh Hoffman

Atomic Tragedy: Henry L. Stimson and the Decision to Use the Bomb against Japan, by Sean Malloy

The Color of War, by James Campbell

Hiroshima in America, by Robert J. Lifton, Greg Mitchell

The Last Wave from Port Chicago, by Peter Vogel

The Practical Einstein: Experiments, Patents, Inventions, by József Illy

Restricted Data: The History of Nuclear Secrecy in the United States, by Alex Wellerstein (University of Chicago Press, 2021).

Target Hiroshima: Deak Parsons and the Creation of the Atom Bomb, by Albert Christman, Naval Institute Press (February 15, 2014)

The New World: A History of the United States Atomic Energy Commission, Volume I. 1939-1946, Richard G. Hewlett and Oscar E. Anderson, Jr. 1990

Articles / Booklets / Pamphlets

- *Against Time: Scheduling, Momentum, and Moral Order at Wartime Los Alamos*, by Charles Thorpe, **Journal of Historical Sociology** Vol. 17 No. 1 March 2004

- *Atomic governance: Militarism, secrecy, and science in post-war America, 1945-1958,* by Mary D. Wammack, University of Nevada, Las Vegas (dissertation)

- *The Big Leak*, by Thomas Fleming, **American Heritage**, December 1987. Volume 38 Issue 8

- *Economist Guided Early Atom Steps*, By Anthony Leviero, Special to the **New York Times**. Nov. 28, 1945

- *Einstein in the U. S. Navy*, by Stephen Brunauer, **Journal of the Washington Academy of Sciences**, Vol. 69, No. 3 (September, 1979), pp. 108-113l;

See also

- *NOTES FROM THE FIELD - Einstein's Ordnance*, Frederic D. Schwarz Spring 1998 (/content/spring-1998-1) | Volume 13 Issue 4 SPRING 1998. **Physics Today**, INVENTION & TECHNOLOGY p 9; Brunauer, S. 1986.

- *Einstein and the Navy: ...an unbeatable combination.* On the Surface. Navy Surface Weapons Centre, Jan 24th at Albert Einstein Archives Online.

- *Einstein's "biggest blunder" – interrogating the legend,* by Cormac O'Raifeartaigh and Simon Mitton; **Physics in Perspective**, 2018 – Springer.

- *Port Chicago Update, History of 10,000 Ton Gadget,* by Harry V. Martin, Copyright FreeAmerica and Harry V. Martin, 1995

See also

Port Chicago - 50 Years: was it an atomic blast? by David Caul and Susan Todd

- *Roosevelt, Truman, and the Atomic Bomb, 1941-1945: A Reinterpretation* by Barton J. Bernstein, **Political Science Quarterly** Vol. 90, No. 1 (Spring, 1975), pp. 23-69 (47 pages)

- *The Rules of Civilized Warfare': Scientists, Soldiers, Civilians, and American Nuclear Targeting, 1940–1945,* By Sean L. Malloy, pp 475-512 Published online: 08 Apr 2008

- *Unspeakable Suffering: – the humanitarian impact of nuclear weapons* - Beatrice Fihn, January 2013; **Reaching Critical Will** – a programme of Women's International League for Peace and Freedom.

More Essays and Posts by Daisy Herndon

For more information and further discussion of the PCnet, including nuclear history, contemporary nuclear policy *and related issues*, follow The Port Chicago Witness Facebook page.

See also

- *The Port Chicago explosion - a nuclear experiment? New evidence of the Manhattan Project connection* (an introduction to my books) on *YouTube*:

- *Uncle Sam's Nuclear Cloak*, four-part **MEDIUM** series by Daisy B. Herndon

1) Uncle Sam's New Cloak, part 1 - what we don't know is killing

2) Uncle Sam's New Cloak, part 2 — race as an issue in World War II

3) Uncle Sam's New Cloak part 3, that time when on one's life mattered

4) Uncle Sam's New Cloak, part 4 — FDR's Big Stick

And visit my Wordpress blog, *The Port Chicago Witness*

- *Exploring the Unexplained Explosion*

- *Background*

- *A Story Too Important to Ignore*

- *Parsons' Priorities*

NOTES

---

[1] FRANKLIN D. ROOSEVELT, 32nd President of the United States: 1933 - 1945, *Undelivered Address Prepared for Jefferson Day*. April 13, 1945. The American Presidency Project.

[2] *Remarks by President Obama and Prime Minister Abe of Japan at Hiroshima Peace Memorial*, Hiroshima Peace Memorial, Hiroshima, Japan, May 27, 2016

[3] *Trinity: "The most significant hazard of the entire Manhattan Project"*, By Kathleen M. Tucker, Robert Alvarez. **Bulletin of the Atomic Scientists**, July 15, 2019

[4] *Destroyer of Worlds : Taking stock of our nuclear present*, by Elaine Scarry, Eric Schlosser, Lydia Millet, Mohammed Hanif, Rachel Bronson, Theodore Postol. **Harper's Magazine**, December 2017

[5] **Half American: THE EPIC STORY OF AFRICAN AMERICANS FIGHTING WORLD WAR II AT HOME AND ABROAD**, By Matthew F. Delmont. Penguin Books, October 2022

[6] *Mississippi Massacre, or Myth? Army Tries to Put to Rest Allegations of 1943 Slaughter of Black Troops* By Roberto Suro and Michael A. Fletcher, **Washington Post** Staff Writers, Thursday, December 23, 1999; Page A04

[7] **Hiroshima: 75 Years Later**, History Channel, History Vault

[8] *Why wasn't Little Boy tested?* By Alan Carr, Editor Whitney Spivey, National Security Science, Los Alamos National Laboratory.

"*The bomb's components were tested, and **Lab scientists** were certain of its success.*" (Emphasis added: Engineers test lab results in the field.)

[9] *Brief History of Radioactivity in metro-St. Louis*, Missouri Coalition for the Environment (MCE), **HAZARDOUS WASTE,** SEPTEMBER 30, 2018

[10] *Let Them Drink Milk*, By Pat Ortmeyer and Arjun Makhijani. IEER (Institute for Energy and Environmental Research) Takoma Park, Maryland, USA October 1997 Updated April 15, 2009

[11] *Nuclear fallout is showing up in U.S. honey, decades after bomb tests.* BY NIKK OGASA, **SCIENCE**, 20 APR 2021

[12] If there is one convenient and apparently unplanned coincidence in the Port Chicago story, it is the prominence of the name "Chicago":

- Camp Robert Smalls was the segregated naval training camp at the Great Lakes Naval Training Center near *Chicago*, where the "Negro" recruits were trained.

- On his way to the West Coast in July 1944, FDR stopped to speak to DNC bosses as they prepared to open the historic political convention in *Chicago*.

- Two days later, on the 17th, when Manhattan Project leaders decided to shelve the Thin Man bomb, they met at the University of *Chicago*.

Many people, upon first hearing of the Port Chicago story automatically assume it refers to Chicago, Illinois. They consequently dismiss the ridiculous (and erroneous) idea that the claim is about a nuclear explosion at "the Port of Chicago".

That, of course, is not the case. That might indeed have been a very hard secret to keep. On the other hand, few people are aware of the "*Baby Crossroads*", the shakedown tests for *Operation Crossroads*, held at San Diego in June 1946, a year after bombing of Hiroshima revealed the secret bomb project.

[13] Barton J. Bernstein, *'It's History – The Quest for an Atomic Torpedo'*, **San Francisco Chronicle**, 5 Aug. 1997, p 19

[14] Lee Basham: "*Governing by Crisis: How Toxic Truths Subvert Mainstream Investigation*", Pitzer College, YouTube, Oct 9, 2017

[15] **Target Hiroshima**, preface, p viii

[16] *A New Nuclear Era Is Coming*, By Uri Friedman, **The Atlantic**, January 9, 2020. *See also*

The Unexpected Return of Duck and Cover, By Glenn Harlan Reynolds, **The Atlantic**, January 4, 2011

Does Nuclear Secrecy Make Us Secure? Alex Wellerstein interview, by Jennifer Ouelette, **Ars Technica**, 4/29/2021

[17] *Unspeakable Suffering: – the humanitarian impact of nuclear weapons* - Beatrice Fihn, January 2013; Reaching Critical Will – a programme of Women's International League for Peace and Freedom.

[18] **Hiroshima in America**, Robert J. Lifton, Greg Mitchell HarperCollins, 1996, p 337.

" **Psychic numbing** can be defined as a diminished capacity or inclination to feel ." (The concept of "psychic numbness", like any theory of social psychology and trauma, is worthy of further investigation, especially in the light of the overarching national and international crises of the early twenty-first century.)

[19] *Inventing a Climate of Opinion: Vannevar Bush and the Decision to Build the Bomb*; Stanley Goldberg; **Isis** Vol. 83, No. 3 (Sep., 1992), pp. 429-452

[20] For Bush's role in the decision to create the bomb, see *The Man Behind the USA's Decision To Build The Bomb,* By Matthew Wills, **JSTOR DAILY**, February 2, 2019

[21] Einstein / Brunauer correspondence, digitized, National Archives

[22] "*Three Who Sold the Atom to America*". Sachs, Einstein, Szilard, from "The Atomic Future", Howard Blakeslee, *Associated Press* Supplement, *Spartanburg Herald*, May 1946, in the papers of Alexander Sachs, p 125

[23]See correspondence between Brunauer and Einstein, stored digitally at the National Archives, in which he discusses 1) use of a torpedo to destroy the hull of a ship; 2) need for experiments due to limitations of mathematical computation; 3) how to position a torpedo to explode parallel to a ship; 4) positioning a torpedo beneath a ship; and related topics.

[24] There is no (apparent) indication that they studied the data from any other large explosions, such as the ones at the Lake Denmark Naval Powder Depot in July 1926; the Yorktown Mine Depot on November 16, 1943; or the West Loch Disaster at the Pearl Harbor Naval base on May 21, 1944. Presumably, none of those disasters had the magnitude of the blast at Port Chicago, which compared to the massive Halifax explosion of 1917 — and to the atomic bomb attack at Hiroshima. For reasons discussed later, the data from Halifax was not useful for the study of the atomic bomb.

[25] *Trinity Tests*, Nuclearweaponarchive.org.

[26] Vogel, "*History of 10000 Ton Gadget Critical Analysis*" -**The Last Wave from Port Chicago, Ch. 6.** See also "*The Authors and the Bomb it describes*" - History of 10000 Ton Gadget, **The Last Wave from Port Chicago** (yumpu.com).

[27] *Talk:Port Chicago disaster/Archive 1 - Wikipedia* and *Talk:Port Chicago disaster - Wikipedia*

[28] Historians and scholars, especially those whose work has been anchored in the traditional history, have been unwilling or unable to investigate the PCnet. For example, Christman insisted that he found nothing in the Parsons' files to substantiate the theory, but this may be most indicative of how a researcher's premises impact their perception. For example, Christman cites the September 1944 letter to Groves where Parsons mentions his "observation of the Port Chicago explosion", but the biographer seems to overlook the most obvious implications of that phrase.

[29]**Target Hiroshima, p 155**

In a postwar speech to the Naval War College on "Problems and Prospects in Atomic Energy", Parsons contradicted Christman's assertion, indicating that the Port Chicago data marked a turning point in the research and development of the bomb.

[30]*SpongeBob SquarePants Theory: Bikini Bottom Is A Nuclear Test Site BY ADRIENNE TYLER, Screenrant.com., PUBLISHED SEP 18, 2019*

[31]*How Kodak Accidentally Discovered Radioactive Fallout;* by Matt Blitz, **Popular Mechanics,** Jun 20, 2016

[32] "*The Atom Bomb as Policy Maker: FDR and the Road Not Taken,*" by Campbell Craig. **The Age of Hiroshima,** p 34

[33] We may have yet to recognize the many ways in which the birth of the nuclear age was indeed the end of an era.

[34] Consodine was deputy to John Lansdale, Groves' chief security and counterintelligence officer.

[35] *Background and Early History, Atomic Bomb Project in Relation to President Roosevelt*, Opening Testimony by Alexander Sachs, **Senate Special Committee on Atomic Energy**, Nov. 27, 1945. See also *Economist Guided Early Atom Steps*, By Anthony Leviero, Special To the **New York Times**. Nov. 28, 1945

[36] Sachs kept copious records. His thick files can be accessed online at the FDR library[1].

[37] *Roosevelt, Truman, and the Atomic Bomb, 1941-1945: A Reinterpretation*: Barton J. Bernstein, **Political Science Quarterly**, Vol. 90, No. 1 (Spring, 1975), pp. 23-69 (47 pages)

1. https://www.fdrlibrary.org/documents/356632/390886/findingaid_sachs.pdf/ab2a7ea0-a007-4d84-b339-a5588366a11f

[38] *Basic Research in the Navy: Report to the Secretary of the Navy* by the Naval Research Advisory Committee · Volume 1 By Arthur D. Little, Inc., United States. Naval Research Advisory Committee · 1959

[39]PBS **History Detectives** feature: *The Bush-Conant files* (pbs.org)

[40] Alexander Sachs Papers, Early Atom Steps - FDR Presidential Library & Museum - "Dr. Alexander Sachs, before Senate group, reveals his first talk with Roosevelt."

[41] *"FDR: I Hardly Know Truman"*, by David McCullough, **American Heritage Magazine** / 1992 / Vol 43, Issue 4

"Thus did Franklin Roosevelt characterize the man who was to be his running mate in 1944 and—as everyone at the astonishing Democratic Convention knew—almost certainly the next President of the United States. Here is FDR at his most devious, Harry Truman at the pivot of his career, and the old party-boss system at its zenith."

[42] *"The Atomic Crusade and Its Social Implications."* Compton, Arthur H. **The Annals of the American Academy of Political and Social Science**, vol. 249, 1947, pp. 9–19. JSTOR.

[43] *"I Hardly Know Truman"* by David McCullough, **American Heritage**, July/August 1992 , Volume 43, Issue 4

[44] *HOW TRUMAN LED THE CHARGE IN CONGRESSIONAL OVERSIGHT*, By ROBERT M. POOLE, **Historynet.com**, 8/1/2012

[45]*FDR, Harry Truman, & the Manhattan Project with Clifton Truman Daniel and FDR Presidential Library.* The National WWII Museum, Jun 24, 2020

[46] *Looking Back: Gen. Marshall and the Atomic Bombing of Japanese Cities*, By Barton J. Bernstein, ARMS CONTROL TODAY, November 2015

[47] *Senator from Missouri, and Democratic Vice-Presidential nominee, Broadcast from Lamar*, Mo., August 31, 1944 **Vital Speeches of the Day**, Vol. X, pp. 706-708.

[48] Quotes in this section are from Joint Chiefs of Staff, Harry S. Truman Library, Folder: *"The Evaluation of the Atomic Bomb as a Military Weapon"* Collection: **The Decision to Drop the Atomic Bomb Series**: President's Secretary's File

[49] **Target Hiroshima**, p 220 (Internet Archive). In Footnote 5, Christman references "Parsons speech, "Results of *Operation Crossroads*," 27 Dec. 1946

[50] **Target Hiroshima**: Ch. 1, A New Kind of Warrior

[51] **Target Hiroshima**: Preface

[52] Paul Williams, **Race, Ethnicity and Nuclear War.** FN 84. Rotter, ***Hiroshima***, pp. 168–72.

[53] *Controversy over the Enola Gay Exhibition*: Bombings of Hiroshima and Nagasaki, Atomic Heritage Foundation, Monday, October 17, 2016

[54]*President Harry Truman Defends Atomic Bombing of Japan as "the Only Language" a "Beast" Can Understand,* Shapell Manuscript Foundation, August 11, 1945

[55] *See* note [v] regarding **Hiroshima in America.**

[56] *President Harry Truman Defends Atomic Bombing of Japan as "the Only Language" a "Beast" Can Understand*, August 11, 1945

[57] Alex Wellerstein, *What journalists should know about the atomic bombings* | **Restricted Data** (nuclearsecrecy.com), June 9th, 2020

[58] Wellerstein, **The Age of Hiroshima**, p 49

[59] *'Rotten', 'Dastardly' Roosevelt Says Of War Charge Made by Wheeler; Phrase 'Plow Under Every Fourth American Boy' He Calls 'Most Unpatriotic Thing That Has Ever Been Said' Roosevelt Turns Ire on Wheeler*; by Frank L. Kluckhohnspecial To the New York Times, Jan. 15, 1941. See also *The Big Leak*, by Thomas Fleming, **American Heritage**, December 1987, Volume 38 Issue 8.

[60] *FDR Meets With Black Leaders*. Side 1, 1637-1972. September 27, 1940. *Transcripts of White House Office Conversations*, 08/22/1940 - 10/10/1940. (FDR had a secret tape recorder running.)

[61] *African Americans threaten march on Washington, 1941*. The Global Nonviolent Action Database, Swarthmore College.

[62] The 1940 MOWM was precursor to the 1963 March on Washington, also led by Randolph, in which Dr. Martin Luther King delivered his famous "I have a dream" speech.

[63] "*This base is going to blow sky high.*" *Port Chicago, 75 years later*, by Public Affairs, UC Berkeley, JULY 17, 2019

[64] **Day of the Bomb: Countdown to Hiroshima** by Dan Kurzman. McGraw-Hill

[65] *WWII - 1945, Tinian Field: Crew of Enola Gay & The Great Artiste Interviewed*. 15 Aug 45, 220644-01. Footage Farm, reel no. 220644-01

[66] **Target Hiroshima**, p161

[67] **Target Hiroshima** p158

[68] Sachs alleged that FDR had agreed to a demonstration, apparently during a private conversation they had in December 1944. There does not appear to be any other source to confirm Sachs' claim. FDR apparently did

not speak to anyone else about a demonstration. The lack of evidence is common, especially with FDR, and is not conclusive one way or another.

The idea that FDR may have momentarily changed his mind should be examined in context of the apparent *lack* of comments about how he looked when he started out on his West Coast trip and reports of his ghastly appearance thereafter. The PCnet takes note of his attack of "the collywobbles" on the morning of the 20th.

[69] *FDR's Jefferson Day speech, April 1945*, prepared for Jefferson Day. Excerpt.

[70] **Target Hiroshima**, p. 194. See also Kurzman, **Day of the Bomb**.

[71] **Target Hiroshima**, p. 112

[72] **Target Hiroshima**, p. 112

[73] Sean Malloy addresses the topic of underwater delivery at some length in pages 58-62 of **Atomic Tragedy: Henry L. Stimson and the Decision to Use the Bomb against Japan**[2]. Malloy identifies several sources in his annotated notes on page 202, including reference to an article by Barton J. Bernstein, "*It's History – The Quest for an Atomic Torpedo*", **San Francisco Chronicle**, 5 Aug. 19, 1997.

Parsons' pursuit of an atomic torpedo was concurrent with Albert Einstein's work on torpedo design as a consultant for the Navy's Bureau of Ordnance.

It also coincided with the development of the *Thin Man* plutonium gun-assembly bomb – a device Parsons once compared to a torpedo.

Parsons to Groves, December 1944, *cited by* Richard Rhodes, *in* **The Making of the Atomic Bomb** p 539

---

2. https://www.google.com/books/edition/Atomic_Tragedy/7uBz0gyy0ysC?hl=en&gbpv=1&dq=nuclear+torpedo+malloy&pg=PA202&printsec=frontcover

"It is believed fair to compare the assembly of the gun gadget to the normal field assembly of a torpedo, as far as mechanical tests are involved. . . ."

[74] **Manhattan District History—Project Y, the Los Alamos Project: Inception until August 1945**, David Hawkins, January 1961, Los Alamos Scientific Laboratory of the University of California

[75] Barton J. Bernstein, "*It's History – The Quest for an Atomic Torpedo*", **The San Francisco Chronicle**, Aug. 5, 1997, p. 19

[76] Bernstein, "*It's History – The Quest for an Atomic Torpedo*", **San Francisco Chronicle**, August 5, 1997, p. 19

[77] *Billy Mitchell and the Battleships* by John T. Correll, **Air &**

Space Forces Magazine, (airforcemag.com.), July 21, 2021; *republished* from 2008

[78] **Target Hiroshima** - Al Christman P. 155

"*In July 1944, with battles raging in Europe and the Pacific, Parsons paced the floor of his office dictating to Hazel Greenbacker his ideas on what form army and navy postwar research should take.*" [Fn 26 Parsons, "Comments on Planning for Army and Navy Research: Difficulties in Peacetime", 8 July 1944, Thompson papers, NLCG. (Navy Laboratories / Center Coording Group collection held by the Naval Historical Center, Washington DC.)

[79] **Target Hiroshima**, p 206

[80] Parsons, W. S. (1948) *Problems and Prospects in Atomic Energy*, **Naval War College Review**: Vol. 1 : No. 3 , Article 2.

[81] *Detonation - Definition, Meaning & Synonyms | Vocabulary.com*

[82] **Target Hiroshima**, p 159. This is *footnote 41.* The document referred to is a memo: "*Special Report of Ordnance and Engineering Activities of Project Y*," Parsons to Groves via Oppenheimer, 25 Sept. 1944, Parsons papers, LC, accessible at the Library of Congress.

[83] Richard Rhodes, **The Making of the Atomic Bomb**, p.548.

(The *History Channel* documentary "Hiroshima at 75" is even more misleading. It says the plutonium would blow apart even before it could be used in a weapon.)

[84]*Los Alamos: Beginning of an Era 1943-1945; Designing the Bomb, Atomicarchive.com, History; Trinity Site: 1945-1995,* A National Historic Landmark, White Sands Missile Range, New Mexico, by White Sands Missile Range Public Affairs Office

[85] The status of the *Thin Man* in July 1944 is not clear. The July 17, 1944 decision to "shelve" it leaves room for interpretation. *See* **The New World: A History of the United States Atomic Energy Commission,** Volume I. 1939-1946, Richard G. Hewlett and Oscar E. Anderson, Jr. 1990. On page 251, Hewlett and Anderson provide insight into the *Thin Man*, the Mark II and Mark I, as discussed in this letter from Conant to Bush.

[86] *The Discovery of Spontaneous Fission in Plutonium during World War II,* by Lillian Hoddeson - **Historical Studies in the Physical and Biological Sciences** Vol. 23, No. 2 (1993), pp. 279-300 (22 pages) University of California Press, Historical Studies in the Physical and Biological Sciences

[87] *Letter of Albert Einstein to Lt. Stephen Brunauer, U.S. Navy Bureau of Ordnance*, *September 1rst, 1943*. In this letter, Einstein discusses the need for experiments.

[88] *A Story Too Good to Kill, The "Nuclear" Explosion in San Francisco Bay,* Badash and Hewlett, **Sage Knowledge**, Volume 14, Issue 4, June 1993.

[89] *Memorandum on Port Chicago Disaster,* W. S. Parsons, July 20, 1944. (atomicarchive.com)

[90] **Target Hiroshima**, p 159

[91] In fact, Port Chicago could not have met the criteria better if the ammunition dump had been *designed* to serve as the first nuclear proving ground. I explore the evidence supporting this even more outlandish extension of the PCnet in my forthcoming book, **Designed for Large Explosions**.

[92] *Edwards Air Force Base,* **Wikipedia**

[93] **Lind may have been** a misspelling; Parsons may have been referring to Maj Gen William E. Lynd, Commanding General, Fourth Air Force, San Francisco, Calif., until 1946. **O**n the other hand, a June 13, 1944 article in the **Oakland Tribune** refers to **Lind**: "commanding general of the fourth air Force (sic) addressed the throng."

[94] That is the subject of my companion book, **Designed for Large Explosions**.

[95]Oliver Hekster, *"The Size of History: Coincidence, Counterfactuality and Questions of Scale in History"* in **The Challenge of Chance**, pp 215-232

"Most historians view coincidences as closely related events that lack causal relationship. That type of coincidence does not fit into a historical narrative, because historians tend to focus on causality, action, and consequence."

[96] *The Ammunition Depots,* **Building the Navy's Bases in World War II**, Vol I, Part II, Chapter 13.

[97] Parsons, W. S. (1948), "*Problems and Prospects in Atomic Energy*," **Naval War College Review**: Vol. 1 : No. 3 , Article 2.

[98] *Report to the President of the National Academy of Sciences* by the Academy Committee on Uranium, November 6, 1941

[99] **Target Hiroshima**, p 109

[100] **Target Hiroshima**, p 111

[101] *CAPABILITIES OF THE ATOMIC BOMB, INCLUDING NAVAL THINKING ON ITS EMPLOYMENT*; W. S. Parsons; **Naval War College Information Service for Officers** , April, 1950, Vol. 2, No. 8 (April, 1950), pp. 23-29; U.S. Naval War College Press

> The sense of military discipline and rigor at Los Alamos in particular is explored by Charles Thorpe in "*Against Time: Scheduling, Momentum, and Moral Order at Wartime Los Alamos*", Journal of Historical Sociology Vol. 17 No. 1 March 2004 ISSN 0952-1909

[102] **Target Hiroshima**, p 204. See also *That Atomic Mushroom,* Rear Admiral W. S. Parsons, **Boys Life**, July 1947, p. 5

[103] *'My God what have we done?': Copy of Hiroshima pilot's log sells for US$50,000"* By AFP, **South China Morning Post**, Published: 9:15am, 30 Apr, 2015

Captain Robert Lewis – co-pilot of the Enola Gay, the bomber that dropped the weapon – wrote in his log "My God what have we done?"

[104] The first of these tests was in 1946 and required the "temporary" evacuation of the islanders, who were told the tests were being conducted "for the good of all mankind". They were never able to return to their homeland, and it is a gross understatement to say that many have suffered severe health problems from overexposure to radioactive fallout.

[105] *The Last Commodore, William S. "Deak" Parsons* by Roger Meade. LA-UR 13-29480; **LANL 70th Anniversary Public Lecture Series**: Los Alamos National Lab, Bradbury Science Museum

[106] *"Problems and Prospects in Atomic Energy,"* **Naval War College,** Parsons, W. S. (1948) **Review**: Vol. 1 : No. 3 , Article 2.

[107] *Nuclear targeting: The first 60 years*; by Arjun Makhijani, **Bulletin of the Atomic Scientists,** May/June 2003 (May 1, 2003)

> "The idea, in May 1943, that atomic bombs could produce a U.S.-dictated world peace was a fantasy then–as it is now."

[108] Sean L. Malloy, "***When You Have to Deal with a Beast": RACE, IDEOLOGY, AND THE DECISION TO USE THE ATOMIC BOMB*,** **The Age of Hiroshima** (pp. 56-70)

[109] "*Answers to Frequent Questions About the Port Chicago Disaster*",

Port Chicago Naval Magazine National Memorial (U.S. National Park Service) (nps.gov) National Park Service

[110] **Guide to Command of Negro Naval Personnel**. NAVPERS-15092, Navy Department, Bureau of Naval Personnel

[111] *Port Chicago disaster*, "*Composition of African American personnel.*" **Wikipedia**, citing *Port Chicago Court of Inquiry, Findings of Facts*

[112] There were two explosions at Port Chicago. In his October 22, 1943 letter to Lt. Stephen Brunauer about using a torpedo to destroy a ship, Albert Einstein recommended two explosions.

[113] The COI cited "Torpex" as an *example* of a 'supersensitive substance', but did not specify it as the probable element that may have caused the explosion.

[114] Mare Island War Diary, July 17, 1944. Fold-3.com

[115] *Trinity test press releases* (May 1945); Alex Wellerstein, **Nuclear Secrecy Blog**, November 10th, 2011

[116] **The Color of War: How One Battle Broke Japan and Another Changed America**, By James Campbell · 2012, pp 65-67

"By the summer of 1943, Naval Ammunition Depots (NADs) were becoming hotbeds of discontent," Campbell wrote, stating, paragraphs later, that "No place was more troubled than Port Chicago."

That characterization seems excessive, particularly in contrast to the racial violence that broke out at some other military bases. This point is taken up in my book, **Designed for Large Explosions**, where we consider why the Navy charged the Port Chicago sailors with conspiracy to conduct mass mutiny, especially when other more flagrant cases did not even go to trial. The book also examines whether or not regulations in force at the time gave enlistees the right to have peers on their court, which was not the case with the Port Chicago 50.

As for the comparison of trouble at military bases, it would require intentional scrutiny to establish similarities and differences. That, of course, is well beyond the scope of the Port Chicago nuclear explosion theory. Interested readers should consult **Half American** by Matthew Delmont, and other books on African-Americans in the military during World War II.

[117] **The Port Chicago Mutiny**, page 53

Quotes in this section are taken from pages 161 – 162

[118] *Port Chicago disaster*, **Wikipedia**.

[119] *Black Scholar Research Leads to Navy Review: Injustice Upheld in Port Chicago Mutiny Trial*, Robert L. Allen**, The Black Scholar** Vol. 24, No. 1, Black Cultural History - 1994 (Winter 1994), pp. 56-59

[120] See, for example,

- *Human Shadow Etched in Stone.* **Wikipedia,** and

- *Fogonazos: Hiroshima, the pictures they didn't want us to see* by Antonio Martínez Ron, **Fogonazos (blog)**, 05 February 2007

[121] The composite photo was *created by the author* from public domain photos of Hiroshima and Port Chicago, respectively, without alteration other than minor cropping. **The sole purpose of the composite photo is to illustrate the comparison made in these paragraphs.**

[122] *Speech of Vice Presidential Candidate Harry S. Truman at Lamar, Missouri, August 31, 1944* – **Truman Library**.

[123] "USS *Baltimore* (CA-68)". Wikipedia

[124] United States Navy, "*The history of the U.S.S. Baltimore, CA-68*" (1946). World War Regimental Histories. 178. (Apparently a commemorative history, published by Bangor Public Library Bangor Community: Digital Commons@bpl World War Regimental Histories - World War Collections)

[125] *The Adventures of Fala, First Dog: The Case of the Dog Who Didn't Bark on the Boat* - **Forward with Roosevelt** - October 12, 2017 Posted in "Found In The Archives", by Paul M. Sparrow, Director, FDR Library.

[126] *AN HISTORIC VOYAGE WITH PRESIDENT OF THE UNITED STATES,. FRANKLIN DELANO ROOSEVELT.* By Douglas E. MacVane, **Forward with Roosevelt**, the blog of the Franklin D. Roosevelt Library and Museum

[127] As MacVane notes later in the essay, FDR's dog, Fala, was indeed along on the trip.

Later, as the campaign heated up, Republican opponents circulated a rumor that the dog had been stranded on an island. They claimed FDR sent a battleship to its rescue. FDR used the occasion to make one of his famous humorous quips, defending his innocent dog. With the one clever remark, he was able to counter talk about the state of his health, reassure his supporters and restore their confidence.

[128] *Baltimore V (CA-68) (navy.mil),* **Naval History and Heritage Command**

[129] MacVane acknowledged that he was writing from memory. A comparison of his story, FDR's itinerary and other primary documents might reconcile the different times given for FDR's arrival at San Diego.

[130] Always the joker, in the draft of his (undelivered) final speech, FDR joked that his doctors could not do anything about his inability to be in two places at one time; it was a chronic condition. Again, this may be a clue to unanswered questions about what FDR knew about his health and when.

[131] Always the joker, in the draft of his (undelivered) final speech, FDR joked that his doctors could not do anything about his inability to be in two places at one time; it was a chronic condition. Again, this may be a clue to unanswered questions about what FDR knew about his health and when.

[132] *With War Looming in Europe in 1939, President Franklin D. Roosevelt Will Personally Witness U.S. Naval War Readiness Exercises - Franklin Roosevelt Typed Letter Signed* | **The Raab Collection**

[133] *FDR's Health* - a virtual tour - FDRLibrary.org

[134] *Why FDR Decided to Run for a Fourth Term Despite Ill Health,* SARAH PRUITT, **History.com**, March 12, 2020

[135] In a letter to first Lady Eleanor Roosevelt, FDR said he had got "the collywobbles—*after* viewing the landing operation. (**Final Victory: FDR's Extraordinary World War II Presidential Campaign**, by Stanley Weintraub pp 2, 3, and 292)

[136] *FDR's Bedside Manner*, **American Heritage**, December 1990 Volume 41 Issue 8

[137] **The Roosevelt myth** by John T. Flynn, January 1, 1948

[138] **Albert Einstein: Creator and Rebel**, by Banesh Hoffman

[139] The torpedo scandal of World War II, along with the consequent scathing declaration by Admiral H. P. Blandy that all new weapons must be tested, weighs against the common assertion that the world's first atomic weapon, the *Little Boy*, was used in combat without having been tested. See also **The Practical Einstein: Experiments, Patents, Inventions**, by József Illy (note 117) and **Iron Men and Tin Fish: The Race to Build a Better Torpedo during World War**, by Anthony Newpower.

[140] "[3] *We had the hose turned on us!": Ross Gunn and the Naval Research Laboratory's early research into nuclear propulsion, 1939-1946*[4]. Joseph - James Ahern. **Historical Studies in the Physical and Biological Sciences**

Vol. 33, No. 2 (2003), pp. 217-236 (20 pages) Published By: University of California Press JSTOR

(NRL's work began almost seven months before President Franklin D. Roosevelt received Albert Einstein's famous letter about the potential for an atomic bomb.)

[141] *Merle Tuve interview* with Albert Christman, 1967 – **American Institute of Physics** (AIP)

3. https://www.jstor.org/stable/10.1525/hsps.2003.33.2.217?read-now=1&refreqid=excelsior%3A9769b845e88d36895549069be9552ff6&seq=4

4. *https://www.jstor.org/stable/10.1525/hsps.2003.33.2.217?read-now=1&refreqid=excelsior%3A9769b845e88d36895549069be9552ff6&seq=4*

"The main thing it did was to provide us a direct channel into the Navy structure, and it gave the Navy confidence. It was a typical thing, a wise thing that Bush did as an administrator. He just saw that this would solve the whole thing for him and for the Navy and for us and for everybody."

Tuve may not have been a reliable witness. It is not clear that he would have had access to the real story. Besides, he also insisted that it hurt Deak Parsons to be the weaponeer on the Enola Gay. In fact, Parsons considered it an awesome responsibility, and was pleased with the delivery. On the other hand, if further research showed that the Navy was indeed funding CIW, the institution led by Bush, that would shed an entirely different light on the idea that the Navy was left out of the Manhattan Project. That, in turn, would merit a review of Einstein's role in the project.

[142] *Cap[5]abilities of the Atomic Bomb, Including Naval Thinking on its Employment* - W. S. Parsons, **Naval War College Review**, April 1950

[143] Coincidentally, Cranbrook is less than 15 minutes from my childhood home; it was one of the schools I might have attended, instead of skipping first grade, but for being labeled emotionally immature. It also seems uncanny that I have lived near, passed through, or had some other connection to several of the sites where Parsons worked, in Maryland, Washington D.C. and Virginia, not to mention my brief stay in the Bay area. To a historian, this might be evidence that coincidences have no intentional cause. From my own spiritual perspective, it is evidence that they do.

Again, I conducted the research and prepared this work as an outlier. As noted from the beginning of this book and demonstrated throughout, there are several reasons why most people who write the topics in this book Port Chicago were not likely to investigate the controversial "conspiracy theory"; even if they had, they would have been guided by prior work and premises, and would have likely missed clues that I picked up precisely because I was not *limited* by such prior knowledge, and because I had no professional standing to lose.

---

5. https://digital-commons.usnwc.edu/cgi/viewcontent.cgi?article=7897&context=nwc-review

I mention these spiritual and emotional topics precisely because they run contrary to the dominant spirit of the American century, the quantifiable, scientific / industrial / material strength—that gave us the bomb.

And then there are intrepid people like Ellsberg, Basham and Vogel, people who believe truth is important, and are committed to telling it, even when it does not comport with our sacred national myths.

[144] Daniel Ellsberg, **The Doomsday Machine: Confessions of a Nuclear War Planner**

[145] *F.D.A., Not F.D.R.* By David Leonhardt, **The New York Times** (nytimes.com) Aug. 24, 2021

[146] *First Inaugural Address of Franklin D. Roosevelt,* Saturday, March 4, 1933. **The Avalon Project**

[147] *"The Evaluation of the Atomic Bomb as a Military Weapon"*, Harry S. Truman (trumanlibrary.gov) (cited above)

[148] *F.D.A., Not F.D.R.* By David Leonhardt, **The New York Times** (nytimes.com) Aug. 24, 2021

"One chose bureaucratic caution. One didn't."

[149] "*Late WWII veteran recalls meeting Einstein*", by Melony Overton, **Port Lavaca Wave**. May 30, 2017 Updated May 30, 2017

[150] The Port Chicago Court of Inquiry found that "The locations of fragments indicate that the explosives in the E. A. BRYAN exploded as one large bomb."

[151] *The Frisch-Peierls Memorandum*, Otto Frisch and Rudolf Peierls:

> "...the bomb could probably not be used without killing large numbers of civilians, and this may make it unsuitable as a weapon

for use by this country. (Use as a depth charge near a naval base suggests itself, but even there it is likely that it would cause great loss of civilian life by flooding and by the radioactive radiations.)

Cited by Sean Malloy in "'*The Rules of Civilized Warfare': Scientists, Soldiers, Civilians, and American Nuclear Targeting, 1940–1945,*[6]" June 2007, **Journal of Strategic Studies** 30(3):475-512

[152] *The Navy in the Manhattan Project* - **Atomic Heritage Foundation**

(The idea that Parsons' investigation of the Port Chicago disaster was part of his study of nuclear attacks on ships seems to confirm the PCnet, as it supports the conclusion (suggested by Christman in *Target Hiroshima*) that *Operation Crossroads* was Parsons' way of tying up a loose end. That "loose end" may have been the disappointing detonation at Port Chicago.)

[153] **Target Hiroshima**, p 113

[154] **Target Hiroshima**, p 113

[155] See *Hearings Regarding Shipment of Atomic Material to the Soviet Union During World War II.* **United States. Congress. House. Un-American Activities** Jan 1950.

(Whether or not FDR or someone in his administration had anything to do with the shipment of uranium to Russia during World War II is beyond the scope of this book. Not much has been written on the subject.)

[156] "*Franklin D Roosevelt – "He gambled on the atom", in "The Atomic Future,"* **Associated Press** supplement, May 1946 (**Spartanburg Herald**) Howard Blakeslee, from Sachs Papers, atomic.07, page 120, FDR Library

[157] **War Without Mercy: Race & Power In the Pacific War**, John W. Dower, W. W. Norton & Company in 1986

6. https://www.tandfonline.com/doi/abs/10.1080/01402390701343482?journalCode=fjss20

[158] *The Belgian Congo at War – "A Primitive Mass of "Natives" Become a Body of Loyal Colonial Citizens"* - Published by the Belgian Information Center 630 Fifth Ave., New York; *Transcribed and formatted* for HTML by Patrick Clancey, HyperWar Foundation

[159] (Craig's title is borrowed from *The Atom Bomb as Policy Maker*, by Bernard Brodie, **Foreign Affairs**, Vol. 27, No. 1 (Oct., 1948), pp. 17-33 (17 pages) : Council on Foreign Relations

[160] Campbell Craig, *The Atom Bomb as Policy Maker: FDR AND THE ROAD NOT TAKEN"* in **The Age of Hiroshima,** (pp. 19-33)

[161] *Capabilities of the Atomic Bomb, Including Naval Thinking on its Employment,* by W. S. Parsons; Naval War College Information Service for Officers , April, 1950, Vol. 2, No. 8 (April, 1950), pp. 23-29; U.S. Naval War College Press

[162] *The Port Chicago disaster and Vallejo*, by BRENDAN RILEY | **Solano County Times-Herald** PUBLISHED: August 26, 2021 at 6:52 p.m. | UPDATED: August 27, 2021 at 12:06 p.m.

[163] Parsons, W. S. (1948) "*Problems and Prospects in Atomic Energy*," *Naval War College Review*: Vol. 1 : No. 3 , Article 2.

[164] Ellsberg, **The Doomsday Machine**, Introduction, p. 13

[165] *The Future of U.S. Nuclear Weapons Policy*, Executive Summary of the National Academy of Sciences Report, Arms Control Association

[166] *Why is America Getting A New $100 Billion Nuclear Weapon?* By Elisabeth Eaves; Art direction by Thomas Gaulkin, **Bulletin of Atomic Scientists**, FEBRUARY 8, 2021

[167] *"Gentlemen, You Are Mad",* Lewis Mumford, **Maclean's,** June 1946

[168] *Capabilities of the Atomic Bomb, Including Naval Thinking on its Employment*, by W. S. Parsons; Naval War College Information Service for Officers , April, 1950, Vol. 2, No. 8 (April, 1950), pp. 23-29; U.S. Naval War College Press

About the Author

Daisy B. Herndon, a former school librarian, was conducting research for a novel set in World War II when she stumbled upon the Port Chicago nuclear explosion theory, which sets the date of the first nuclear explosion back to July 17, 1944 at the Naval Ammunition Depot near San Francisco. Ridiculous? She thought so too, until she noticed the date on the memorandum authorizing the court martial of the Port Chicago 50. The Black sailors were found guilty of mutiny after refusing to load ammunition onto ships for fear of *another unexplained explosion.* But why was the trial set up on ***July 14th, three days before the massive blast*** that "atomized" their fellow sailors? Herndon currently resides somewhere in Maryland and enjoys watching birds, especially with her grandchildren.

Read more at https://theportchicagowitness.wordpress.com/.

www.ingramcontent.com/pod-product-compliance
Lightning Source LLC
LaVergne TN
LVHW052013080426
835513LV00018B/2022
9798986053615